DNA Methylation Microarrays

Experimental Design and Statistical Analysis

Chapman & Hall/CRC Biostatistics Series

Editor-in-Chief

Shein-Chung Chow, Ph.D.
Professor
Department of Biostatistics and Bioinformatics
Duke University School of Medicine
Durham, North Carolina, U.S.A.

Series Editors

Byron Jones
Senior Director
Statistical Research and Consulting Centre
(IPC 193)
Pfizer Global Research and Development
Sandwich, Kent, UK

Jen-pei Liu
Professor
Division of Biometry
Department of Agronomy
National Taiwan University
Taipei, Taiwan

Karl E. Peace
Director, Karl E. Peace Center for Biostatistics
Professor of Biostatistics
Georgia Cancer Coalition
Distinguished Cancer Scholar
Georgia Southern University, Statesboro, GA

Bruce W. Turnbull
Professor
School of Operations Research
and Industrial Engineering
Cornell University
Ithaca, NY

CH Chapman & Hall/CRC Biostatistics Series

Published Titles

1. *Design and Analysis of Animal Studies in Pharmaceutical Development,* Shein-Chung Chow and Jen-pei Liu
2. *Basic Statistics and Pharmaceutical Statistical Applications,* James E. De Muth
3. *Design and Analysis of Bioavailability and Bioequivalence Studies, Second Edition, Revised and Expanded,* Shein-Chung Chow and Jen-pei Liu
4. *Meta-Analysis in Medicine and Health Policy,* Dalene K. Stangl and Donald A. Berry
5. *Generalized Linear Models: A Bayesian Perspective,* Dipak K. Dey, Sujit K. Ghosh, and Bani K. Mallick
6. *Difference Equations with Public Health Applications,* Lemuel A. Moyé and Asha Seth Kapadia
7. *Medical Biostatistics,* Abhaya Indrayan and Sanjeev B. Sarmukaddam
8. *Statistical Methods for Clinical Trials,* Mark X. Norleans
9. *Causal Analysis in Biomedicine and Epidemiology: Based on Minimal Sufficient Causation,* Mikel Aickin
10. *Statistics in Drug Research: Methodologies and Recent Developments,* Shein-Chung Chow and Jun Shao
11. *Sample Size Calculations in Clinical Research,* Shein-Chung Chow, Jun Shao, and Hansheng Wang
12. *Applied Statistical Design for the Researcher,* Daryl S. Paulson
13. *Advances in Clinical Trial Biostatistics,* Nancy L. Geller
14. *Statistics in the Pharmaceutical Industry, Third Edition,* Ralph Buncher and Jia-Yeong Tsay
15. *DNA Microarrays and Related Genomics Techniques: Design, Analysis, and Interpretation of Experiments,* David B. Allsion, Grier P. Page, T. Mark Beasley, and Jode W. Edwards
16. *Basic Statistics and Pharmaceutical Statistical Applications, Second Edition,* James E. De Muth
17. *Adaptive Design Methods in Clinical Trials,* Shein-Chung Chow and Mark Chang
18. *Handbook of Regression and Modeling: Applications for the Clinical and Pharmaceutical Industries,* Daryl S. Paulson
19. *Statistical Design and Analysis of Stability Studies,* Shein-Chung Chow
20. *Sample Size Calculations in Clinical Research, Second Edition,* Shein-Chung Chow, Jun Shao, and Hansheng Wang
21. *Elementary Bayesian Biostatistics,* Lemuel A. Moyé
22. *Adaptive Design Theory and Implementation Using SAS and R,* Mark Chang
23. *Computational Pharmacokinetics,* Anders Källén
24. *Computational Methods in Biomedical Research,* Ravindra Khattree and Dayanand N. Naik
25. *Medical Biostatistics, Second Edition,* A. Indrayan
26. *DNA Methylation Microarrays: Experimental Design and Statistical Analysis,* Sun-Chong Wang and Arturas Petronis

Chapman & Hall/CRC Biostatistics Series

DNA Methylation Microarrays

Experimental Design and Statistical Analysis

Sun-Chong Wang

Arturas Petronis

CRC Press
Taylor & Francis Group
Boca Raton London New York

CRC Press is an imprint of the
Taylor & Francis Group, an **informa** business
A CHAPMAN & HALL BOOK

CRC Press
Taylor & Francis Group
6000 Broken Sound Parkway NW, Suite 300
Boca Raton, FL 33487-2742

First issued in paperback 2019

© 2008 by Taylor & Francis Group, LLC
CRC Press is an imprint of Taylor & Francis Group, an Informa business

No claim to original U.S. Government works

ISBN-13: 978-1-4200-6727-9 (hbk)
ISBN-13: 978-0-367-38740-2 (pbk)

Library of Congress Cataloging-in-Publication Data

Wang, Sun-Chong, 1963-
DNA methylation microarrays : experimental design and statistical analysis / author/editors, Sun-Chong Wang and Art Petronis.
p. cm. -- (Chapman & Hall/CRC Biostatistics series)
Includes bibliographical references and index.
ISBN 978-1-4200-6727-9 (alk. paper)
1. DNA microarrays--Statistical methods. 2. DNA microarrays--Research--Methodology. 3. DNA--Methylation--Statistical methods. 4. DNA--Methylation--Research--Methodology. I. Petronis, Art. II. Title. III. Series.

QP624.5.D726W36 2008
572.8'636--dc22   2007052279

Visit the Taylor & Francis Web site at
http://www.taylorandfrancis.com

and the CRC Press Web site at
http://www.crcpress.com

Preface

This book is intended to help researchers and students analyze high through-put epigenomic data, in particular DNA methylation microarray data, with sound statistics. It is divided into four parts. Part I, chapters 1 to 4, begins with an introduction to the basic statistics that is needed to comprehend the rest of the book. Chapters 2 and 3 are more biological and describe the wet bench technologies producing the data for analysis. Bioinformaticians and experimenters benefit by working closely together to design experiments with sufficient power and high quality. Chapter 4 preprocesses the data to remove systematic artifacts resulting from imperfection in the measurement. In Part II, chapters 5 and 6, the normalized data are then subject to conventional hypothesis-driven analysis looking for differential methylated loci between populations. Genomic tiling arrays may differ in their specific aims and a whole chapter is devoted to tiling arrays. Part III, chapters 7 to 10, concerns exploratory analysis, which hopefully lead us to new hypotheses. In particular, functions and roles of unannotated DNA elements are associated with those of known ones by cluster and network analysis. Part IV, chapters 11 to 13, introduces the public online resources that facilitate justification and discussion of the findings.

DNA methylation microarrays share the same underlying hybridization principles as gene expression microarrays. Many of the analyses in the book, therefore, apply not only to DNA methylation but also to gene expression and histone modifications by chromatin immunoprecipitation on chip.

Graphical and artistic presentation of the data is no less crucial than statistical computation. We strove to produce decent plots that serve as examples of the resulting analysis throughout the chapter. Electronic files of the plots accompany the book. The data used to illustrate the book were produced at the Krembil Family Epigenetics Laboratory, Centre for Addiction and Mental Health in Toronto, Canada.

We are indebted to the entire team of the KFEL, especially Axel Schumacher, James Flanagan, Jonathan Mill, Zachary Kaminsky, Thomas Tang, Carolyn Ptak, and Gabriel Oh for enlightening discussions and comments. Part of the work was supported by grants from the National Science Council (Taiwan) to Sun-Chong Wang and from the National Institute of Mental Health (R01 MH074127), Canadian Institutes for Health and Research, Ontario Mental Health Foundation, NARSAD, and the Stanley Foundation to Arturas Petronis.

Sun-Chong Wang
Systems Biology and Bioinformatics Institute
National Central University
Chungli Taoyuan, Taiwan

Arturas Petronis
Epigenetics Laboratory
Centre for Addiction and Mental Health
Toronto, Ontario, Canada

Contents

List of Tables

List of Figures
(See accompanying CD for figures featuring color)

Chapter 1

Applied Statistics

Phenotype is shaped by DNA sequence variation, environment, epigenetic variation and their interactions. A few measures in statistics help summarize the characteristics in the collected data regarding the location and shape of the data distribution. Graphical presentation of data is especially useful in comprehending, interpreting and contrasting results, such as the relations between measured quantities.

Description of data changes from one to another independent sample because of the inherent biological variability. Experimental goals are, however, invariantly aimed at the interests of the general population. Methods are needed for us to generalize the results beyond local sample. The central limit theorem that relates sample characteristics to population characteristics underpins the method of induction. Statistical hypothesis tests are introduced to control the risk of false claims. Tests differ depending on the way of comparisons and/or on the type of data.

Relations among data can be explored by the techniques of linear models and contingency tables. We address the sample size issue, which is essential to project planning. Equations are provided for deductive reasoning; most statistical software/packages, such as R/Bioconductor, encapsulate the formulas into single functions, returning results after a mouse click on a user interface. The job becomes that of choosing the right options to the right functions. Chapter 1 serves this purpose.

1.1 Descriptive statistics

Given N independent measurements, it is often desirable to summarize the large number of data values in ways that are succinct yet informative. Examples are abundant, and mundane examples include the blood pressure measured in N days and the time spent on the road to the office in N trips. The quantity of primary interest is the typical value in the sample data. With such typical values, one gets a sense of how effective the medication is she has been receiving and how soon she has to leave for the office. The other quantity of essential importance is the dispersion in the data values. On days of blood pressure highs, she isn't panicking because of the variability in the readings. On occasion of important dates, she is better off leaving early for the meeting. The two quantities in the data values, i.e., central tendency and variation, therefore, are of central importance and we use them daily without notice.

There are measures of central tendency and variation. Different measures can give rise to quite different values. Preferred choice for measures depends on the context. A pictorial presentation of the data helps guide one through the selection of the appropriate measure.

1.1.1   Frequency distribution

The most common way of visualizing a dataset is to group the data values into different intervals or categories. For example, we count the numbers of times the blood pressure (traveling time) falls within the interval between x and $x + \delta x$, between $x + \delta$ and $x + 2\delta$, ..., etc, mmHg (hr). Next we plot the counts over the whole range of the measured values and get a frequency distribution of the data. Note that since the total number of measurements is N, the sum of all the counts in the frequency distribution is equal to N. This kind of a plot is called a histogram. Since the process of making histograms is so routine in statistical data analysis, the word histogram is being used as a verb.

When we compare two frequency distributions, the comparison can be difficult when the two Ns differ by a lot. In such cases, we normalize the counts by dividing them by N, the sum of all the normalized counts now becoming one. The normalized frequency distributions now can be easily compared because the areas under the normalized counts are all the same. Figure 1.1 shows an example of the probability density distribution. The other use of normalized frequency distributions is that, when N is large enough, they can be interpreted as probability density distributions, which is discussed below.

1.1.2   Central tendency and variability

A measure for central tendency in the data values $\mathbf{x} = x_i, x_2, \cdots, x_N$ is the sample average or arithmetic mean, $\bar{x}$,

$$\bar{x} = \frac{1}{N} \sum_{i=1}^{N} x_i \, . \tag{1.1}$$

The mean value (1.1) is prone to shift by extreme values in the data distribution. Extreme values can occur due to the nature of the processes generating the data or to insufficient numbers of data values. For example, the commuter time is protracted by a subway strike. If the increased time happens to be included in dataset $\mathbf{x}$ in mean calculation, the obtained average is inflated by the outlying value.

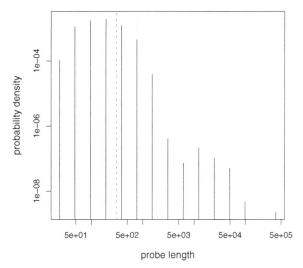

FIGURE 1.1: Length distribution of the probes on a CpG island microarray. The dotted line locates the average (= 773 bp) while the dashed line is the median length (= 306 bp).

Median is a way to remedy the nonrobustness caused by extreme values. To calculate the median, we sort the data values in increasing order. The value in the middle of the sorted data string is the median. In case there is an even number of data values, the average of the two middle values is used as the median. Median is the second quartile in the data distribution; half of the data values are smaller and the other half are larger. Because it is rank that is involved in locating median, it is clear that median is robust to outliers.

Means or medians give an idea of the typical magnitude representing the bunch of data values. As important is the dispersion in the values, which is quantified by the standard deviation,

$$s = \sqrt{\frac{1}{N-1} \sum_{i=i}^{N} (x_i - \bar{x})^2} \ . \tag{1.2}$$

The summation inside the square root is over squared deviations from the mean. $N - 1$ in the denominator is equal to the number of independent data values for the calculation of s. That is, given $\bar{x}$, only $N - 1$ pieces of independent information remain in the dataset. It is called degrees of freedom in statistics.

Like mean, standard deviation can vary from dataset to dataset if some datasets contain outliers. A measure of variability that is resistant to outliers is MAD (median absolute deviation) using medians,

$$\text{MAD} = \text{median}\big(|x_i - \text{median}(\mathbf{x})|\big) \ . \tag{1.3}$$

The other common robust measure for distributional dispersion is IQR (interquartile range), which is the difference between the third and first quartiles of the data values.

1.1.3   Correlation

In section 1.1.1, we advocated graphical presentation of the data with frequency distributions prior to analysis. From the distributional curves in the plot, we decide not only suitable measures of central tendency and variability for the description of the data, but also the type of correlation and test for subsequent analysis.

1.1.3.1   Pearson product-moment correlation

If we position diet consumption according to the scale on the horizontal axis and the blood pressure of the same day along the vertical axis, we form a graph called scatter plot. In such a plot of consumption versus pressure, if the pressure increases linearly with consumption, we say pressure is positively correlated with consumption. Likewise, if blood pressure decreases linearly with drug dosage, the pressure is said to negatively correlate with the drug. We call properties (such as blood pressure or different dosages) that vary between individuals or over time *variables*. The variation may be due to randomness, treatment or free will. The variables that are measured are called dependent or response variables and those that are manipulated are called independent variables, explanatory variables or factors. A quantity for the strength and direction of the linear relation between two variables x, y is the Pearson product-moment correlation coefficient (after Karl Pearson),

$$r = \frac{\sum_{i=1}^{N}(x_i - \bar{x})(y_i - \bar{y})}{(N-1)s_x s_y} , \qquad (1.4)$$

where $\bar{x}$, $\bar{y}$ and s_x, s_y are the sample means and standard deviations for x and y, respectively.

Pearson correlation coefficient is between 1 and -1. A value of 0 indicates no tendency of co-variation between the two variables. In social sciences, a coefficient of 0.3 (or -0.3) is considered large, warranting further investigation. Figure 1.2 shows examples of extreme correlation from technical replicates. Note that correlation does not mean causation. For example, the increasing blood pressure might have been caused by some other environmental factors, such as stress, which have not been taken into account in the analysis. A cause-and-effect relationship between two variables is hard to prove, if not impossible. A way of approximating a proof is to rule out as many false causes as possible.

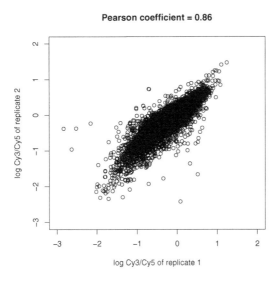

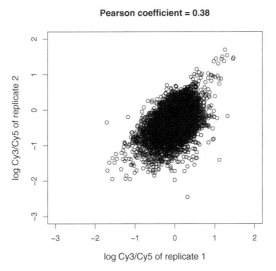

FIGURE 1.2: Scatter plot of replicated DNA methylation microarray measurements. Top shows two replicates of the tissue sample from an individual. Bottom shows two replicates of the tissue from two individuals.

1.1.3.2   Spearman's rank correlation

If the variable increases monotonically, but not linearly, with the other variable in a scatter plot, Pearson correlation coefficient can return a misleading result. In such cases where linearity may not be assumed, we calculate instead Spearman's rank correlation coefficient (after Charles Spearman). To get Spearman's correlation coefficient, we convert the (usually continuous) data values into ordinal numbers according to their rank in relation to others. The rankings are then fed into equation (1.4), producing Spearman's rank correlation coefficient. The coefficient obtained this way, therefore, quantifies the degree of co-variation between the ranks of one variable and those of the other variable.

1.2   Inferential statistics

Finite time and resources bound the number of subjects for a research. The measures introduced in descriptive statistics characterize the distributions of phenotypes pertaining to the participating subjects. The goal of most research, however, is to extend the conclusion on the studied subjects to the general public. Making inferences beyond what is described in the sample data is not possible without probability theory and implicit assumptions [Casella01, Hogg05].

1.2.1   Probability distribution

Probability is used to deal with uncertainty, which originates from the stochastic and/or random nature of the variable. The blood pressure of an individual changes with her physiological conditions around the clock. If we reduce the intraindividual variability by taking blood pressure before, say, morning meals, blood pressure still varies across individuals over the variant genetic backgrounds.

Suppose a procedure against high blood pressure is being developed and that the average blood pressure of N patients is measured for the purpose of efficacy assessment. If we recruit another independent set of N patients, the average blood pressure of the new set will be different from the previous average because of the variability in between-individual blood pressure. Averages differ from set to set, giving rise to a distribution of the means. It is important to know how precise the mean of a sample set is representing the population mean before we can proceed with the assessment.

A probability distribution of the blood pressure $P(y)$ would help by telling us the likelihood of picking an individual with a particular blood pressure y if she is picked at random from the population. The mean and standard deviation of the population blood pressure is not, and will probably never be, known. Thus, the question is how well the population mean is approximated by a sample mean. Central limit theorem holds the key.

1.2.2 Central limit theorem and normal distribution

The central limit theorem is the foundation of many statistical methods. The mathematically proven theorem says that, from a set of N independent variables $y_1, y_2, \cdots, y_N$ each of which has a finite mean μ and standard deviation σ, the following mean

$$\bar{Y} = \frac{y_1 + y_2 + \cdots + y_N}{N} \tag{1.5}$$

is normally distributed with mean μ and standard deviation $\sigma/\sqrt{N}$ as N is large (e.g., $N > 30$).

In our example, μ and σ are the mean and standard deviation of the population blood pressure distribution $P(y)$. The central limit theorem says that the sample mean $\bar{Y}$ approaches the population mean μ for large N. Furthermore, the larger the N, the better the approximation as the standard deviation of $\bar{Y}$ is proportional to $1/\sqrt{N}$.

Central limit theorem also explains the ubiquity of normal distributions. As a quantitative trait with continuous levels, blood pressure can be thought to be under the influence of n genes. Imagine that the allelic effect of gene i on men's blood pressure is z_i. An individual's blood pressure y is then $y = z_1 + z_2 + \cdots + z_n$. If we apply the central limit theorem to y, the probability distribution $P(y)$ of an individual's pressure follows a normal distribution.

The central limit theorem also turns out to be a windfall to statisticians because Gaussian functions are known to be tractable mathematically. Many powerful tools in statistics, to be introduced below, thus have been developed assuming the data are normally distributed. If not, transformations of the data into normal are tried so that existing tools can be applied.

1.2.3 Statistical hypothesis testing

We readily get the mean from a sample of N patients. Although the sample mean approximates the population mean, it does fluctuate. To claim the efficacy of a treatment, which was tested on a limited number of sample patients, to all the patients in a population, we run into the risk of making false efficacious claims. The procedure of statistical hypothesis testing helps us minimize the risk.

1.2.3.1 Null hypothesis

First of all, efficacy itself can be hard to define. For example, a reduction in the average blood pressure by $\bar{Y}$ mmHg can be beneficial to some but not to others. Instead, we usually put forward a null hypothesis H_o of no effect on the average blood pressure after treatment, which is easy to state qualitatively and quantitatively. Negation of the null hypothesis is the alternative hypothesis H_a, which is usually what researchers are looking after. It is good to be reminded that, although we use data from the sample, the conclusion is to be inferred to the larger patient population.

The aim of biomedical research eventually comes down to a null hypothesis. A null hypothesis also dictates what data are collected and compared. In the drug efficacy example, the data are the blood pressure of each patient before and after treatment. We, therefore, form N pre- minus post-treatment blood pressure differences from the N patients. Suppose we find an average difference of $\bar{Y}$ mmHg in the N patients. Recall that the blood pressure fluctuates irrespective of treatment. So what do we decide on the effect of the drug on high blood pressure patients at large?

1.2.3.2 Student's *t*-distribution and p-value

The idea is to have the distribution of the average difference $\bar{Y}$ so that we know how wildly it can fluctuate. We, therefore, calculate the so-called *t*-statistic,

$$t = \frac{\bar{Y}}{s_Y/\sqrt{N}} , \tag{1.6}$$

which, under the null hypothesis of no population difference $H_o : Y = 0$, is shown to distribute as a probability density function called Student's *t*-distribution (after William Sealy Gosset) with $N-1$ degrees of freedom. The s_Y in the denominator of equation (1.6) is the sample standard deviation of the N blood pressure differences using equation (1.2). With the *t*-distribution and the degrees of freedom, we know how $\bar{Y}$ behaves. In particular, given a value of t by equation (1.6), we calculate the area, under the *t*-distributional curve, larger than t. Since the *t*-distributional curve is normalized to 1, the area is the probability of observing a blood pressure difference larger than $\bar{Y}$.

To make conclusions based on the sample data, we estimate the probability of observing the difference or larger assuming that that drug has no effect. The probability, called (one-tailed) p-value, is thus the chance that we get a difference larger than $\bar{Y}$ by chance alone. If the p-value of $\bar{Y}$ is as large as, say 0.5, then the drug effect is not significant because we expect to observe such a $\bar{Y}$ once every two such assessments given that the drug is not effective. On the other hand, if the p-value is as small as, say 0.05, then because the chance of getting such a $\bar{Y}$ by pure chance is only 1 out of 20, we are willing to bet that this blood pressure difference $\bar{Y}$ is due to the drug. The p-value of the hypothesis test indicates the statistical significance of the test result.

1.2.3.3 Type I and type II errors

In the above, we used 0.05 as the significance level, $\alpha = 0.05$, to reject our null hypothesis, embracing the alternative hypothesis that the drug had an effect. The level is the chance that the drug did not have any effect while we erroneously claim so. It is called type I error or false positive. We usually set the level as low as possible so that we do not commit this type of error.

A low level of significance indicates that we attribute the observed blood pressure difference to a small chance rather than to a drug. If the drug did work, we fail to discover its effect and commit a type II error or false negative.

The chance of rejecting a null hypothesis, which is indeed false, is equal to one minus type II error rate and called statistical power or, simply, power. It is seen that the lower the level of significance set for the test, the lower its power. A significance level of 0.05 is chosen by most researchers for the tradeoff between type I and type II errors.

1.2.4 Two-sample *t*-test

The blood pressure data in our example are special in the sense that a patient's blood pressure before the treatment is compared to that after the treatment; the data from a patient are matched or paired. The average of the pretest minus posttest differences is tested for (or against) the null hypothesis of no difference $H_o : Y_1 = Y_2$ where Y_1 and Y_2 are the means of populations 1 and 2. Many other applications, such as comparing the blood pressure between a healthy male and a healthy female, allow no pairing. In these cases, which are more often encountered, we again calculate the *t*-statistic,

$$t = \frac{\bar{Y}_1 - \bar{Y}_2}{\sqrt{(s_1^2 + s_2^2)/N}} \,, \tag{1.7}$$

where $\bar{Y}_i$ and s_i are the sample means and standard deviations of group $i = 1, 2$ and N is the number of individuals per group. The *t*-statistic follows the Student's *t*-distribution with $2N - 2$ degrees of freedom. The sample size per group can be unequal: $N_1 \neq N_2$. The denominator of equation (1.7), called standard error, needs to be slightly modified, so does the degrees of freedom when $N_1 \neq N_2$.

1.2.5 Nonparametric test

Distribution of the data values in the sample is assumed normal in the derivation of Student's *t*-distribution for the significance of the test statistics equation (1.6) and equation (1.7). The normality condition does not always hold in the sample data, as outliers may occur. When no distributional shape in the data is assumed, we use the tests proposed by Frank Wilcoxon for the comparison of distribution locations.

1.2.5.1   Wilcoxon signed-rank test

Without regard to distributional shape of the data, if the treatment has no effect on blood pressure, we expect that half of the pretest−posttest differences are positive and half are negative. That is to say that the median of the differences is zero. We can use the Wilcoxon signed-rank test to examine the null hypothesis of zero median in the matched differences,

$$H_o : \text{median}(Y) = 0 , \tag{1.8}$$

where Ys are patients' blood-pressure differences before and after treatment. To estimate the significance of the median difference, a test statistic is formed involving the sum of signed ranks of the absolute differences. The distribution of the test statistic is tabulated (or approximated by a normal distribution when $N > 10$) from which associated p-values can be estimated. Wilcoxon signed-rank test can also be applied to ordinal data. It is a nonparametric alternative to the paired Student's t-test.

1.2.5.2   Wilcoxon rank-sum test

For unmatched data where strict normality is not true, we can test the following null hypothesis using Wilcoxon rank-sum test,

$$H_o : \text{median}(Y_{1_i} - Y_{2_j}) = 0 , \tag{1.9}$$

where Y_{1_i} is the ith data value from population 1 and Y_{2_j} is the jth data value from population 2 with i and j running over the members of each population. The test involves sorting the combined data into ranks. The sum of the ranks from one group, the statistic in this test, should be close to that from the other group if the two data distributions are from a single population. The p-values are calculated from the distribution of the test statistic. Wilcoxon rank-sum test is equivalent to a two-sample t-test on rankings of the combined data.

Because mean values are sensitive to outliers, if the data distribution is not symmetric, we suggest Wilcoxon tests lest the t-tests report spurious significance. When the data distributions are not far from normal, parametric tests of equation (1.6) and equation (1.7) are more efficient (i.e., having higher power). We may bring nonnormal data into more or less normal by transformation and employ the parametric t-tests for mean comparison.

1.2.6   One-factor ANOVA and F-test

When experimentation gets so involved that more than two groups (e.g., treatments) are compared, we can, in principle, apply two-sample t-tests to each of the pair of group means. For k groups, the number of possible pairs is $k \times (k-1)/2$, 5 percent of which are expected to be false positives. We introduce the F-test that improves over the multiple t-tests by analysis of variance (ANOVA).

Suppose we recruit $n = k \times N$ independent patients for treatment comparison. We randomly partition the patients into k groups, each of which is assigned a treatment. Note that one of the groups receives a placebo as control. The experimental objective is to test any difference in the patients' responses to the various treatments. We set up the null hypothesis, which in the setting reads,

$$H_o : \mu_1 = \mu_2 = \cdots = \mu_k , \tag{1.10}$$

where μ_i is the mean response of patients of population i. Recall again that the hypothesis refers to populations.

If the null hypothesis is true, the sample means $\bar{Y}$s are just various realizations of the same population and from the central limit theorem the $\bar{Y}$s distribute as a normal distribution with variance σ^2/k, where σ is the standard deviation of the population. The variance of $\bar{Y}$ can be estimated by MS_B,

$$MS_B = \frac{SS_B}{k-1} = \frac{N \sum_{i=1}^{k}(\bar{Y}_i - \bar{\bar{Y}})^2}{k-1} , \tag{1.11}$$

where $\bar{\bar{Y}}$ is the mean of the means $\bar{Y}$s. Note that each deviation of $\bar{Y}$ from $\bar{\bar{Y}}$ is weighted by the number of subjects in the group. The variance under the null hypothesis σ^2/k can be estimated by MS_W,

$$MS_W = \frac{SS_W}{k(N-1)} = \frac{\sum_{i=1}^{k} \sum_{j=1}^{N}(Y_{ij} - \bar{Y}_i)^2}{n-k} , \tag{1.12}$$

where Y_{ij} is the response of patient j in group i. If the null hypothesis is true, the found variance MS_B should be close to the expected variance MS_W, i.e., the F-statistic,

$$F = \frac{MS_B}{MS_W} , \tag{1.13}$$

should be $\simeq 1$. The distribution of F-statistic was shown to follow the F-distribution (in honor of Sir Ronald Aylmer Fisher) with $k-1$ and $n-k$ degrees of freedom under the null hypothesis. Given a level of significance $\alpha = 0.05$, we determine the p-value of the above F-ratio. If the p-value is smaller than 0.05, the data do not support the null hypothesis, suggesting that at least one of the equality in equation (1.10) does not hold.

1.2.7 Simple linear regression

In addition to tests for differences in means, we may desire to explain the observed data by constructing a model like,

$$y = \beta_0 + \beta_1 x + \epsilon , \tag{1.14}$$

where y is the response variable, β_0 an intercept term, x the explanatory variable, β_1 the corresponding parameter (or regression coefficient) and ϵ an error term to account for anything that is not taken care of by the parameters. The response y can be blood pressure and the x weight. The model (1.14), therefore, predicts β_0 units of pressure when the weight is zero and a change of β_1 units of pressure per unit increase in weight. The model (1.14) is called linear regression because it is linear in the parameters βs. Now suppose we have collected N independent sets of data, each set recording an individual's blood pressure and weight. We can estimate the two parameters from the data assuming the ϵs for the blood pressures from individuals are independent and identically distributed (i.i.d.) as a normal distribution with mean zero and variance σ^2.

To test the model (1.14), we form the null hypothesis,

$$H_o : \beta_1 = 0 \ , \tag{1.15}$$

asserting the blood pressure does not change with weight. If we graph the N pairs of y and x in a scatter plot, the model (1.14) describes a straight line with slope and intercept β_1 and β_0. The parameters for the population are unknown. We can estimate them from the data using the method of ordinary least squares, which amounts to tuning the slope and intercept so that the sum of the squared distances of the data points to the line is minimized. The least squares estimates for the population parameters are

$$b_1 = \frac{\sum_{i=1}^{N}(x_i - \bar{x})(y_i - \bar{y})}{\sum_{i=1}^{N}(x_i - \bar{x})^2} \ , \tag{1.16}$$

for the slope and $b_0 = \bar{y} - b_1\bar{x}$ for the intercept. With b_0 and b_1, we can estimate y given any x: $\hat{y}_i = b_0 + b_1 x_i$, $i = 1, 2, 3, \cdots, N$. The least squares estimate of b_1 is associated with uncertainty because of the random sampling of y. This is seen from the term $y_i - \bar{y}$ in equation (1.16). The variance of b_1 is $\sum_i (\partial b_1/\partial y_i)^2 \sigma^2$ according to how errors propagate. The standard error of b_1 is, therefore, shown to be

$$SE(b_1) = \sqrt{\frac{\sum_{i=1}^{N}(y_i - \hat{y})^2/(N-2)}{\sum_{i=i}^{N}(x_i - \bar{x})^2}} \ . \tag{1.17}$$

Under the null hypothesis of zero slope, the t-statistic,

$$t = \frac{b_1}{SE(b_1)} \ , \tag{1.18}$$

follows a t-distribution of $N - 2$ degrees of freedom, from which we calculate the one-tailed p-value giving us the probability for the data to generate a larger than b_1 slope by chance when, in fact, the slope in the population is

TABLE 1.1: A Contingency Table for Chi-Square Test

BMI province	A	B	C	D	Any Province
Underweight	uA	uB	uC	uD	uA+uB+uC+uD
Normal	nA	nB	nC	nD	nA+nB+nC+nD
Overweight	oA	oB	oC	oD	oA+oB+oC+oD
Any weight	uA+ nA+ oA	uB+ nB+ oB	uC+ nC+ oC	uD+ nD+ oD	Total

zero. Notice the N in $SE(b_1)$ in equation (1.17). To lower the uncertainty in slope determination, one way is to increase the sample size N.

Recall the equivalence of t-test to F-test: $t_{df}^2 = F_{1,df}$. If we square equation (1.18), it can be shown that,

$$F = \frac{\sum_{i=1}^{N}(\hat{y}_i - \bar{y})^2}{\sum_{i=1}^{N}(y_i - \hat{y}_i)^2/(N-2)} = \frac{SS_M}{SS_E/(N-2)} = \frac{(SS_T - SS_E)}{SS_E/(N-2)}, \quad (1.19)$$

where $SS_T = \sum_{i=1}^{N}(y_i - \bar{y})^2$ is the total sum of squares. Since SS_T is associated with $\beta_1 = 0$ and SS_E with $\beta_1 \neq 0$, the last equality tells us that the F-statistic of equation (1.19) is measuring the reduction in the total sum of squares when β_1 is turned on. If the improvement is large enough, we gain support for the regression model (1.14).

The model (1.14) can be readily extended to include more than one independent variable. It is worth noting that the correlation of equation (1.4) is related to the linear model (1.14) in the way,

$$r^2 = \frac{SS_M}{SS_T}. \quad (1.20)$$

A interpretation of r^2, therefore, says that it measures the fraction of variability in the dependent variable that can be explained by the model (i.e. independent variable(s)). It is called coefficient of determination in statistics. In section 5.8, we illustrate a nonparametric permutation method to assess the statistical significance of r.

1.2.8 Chi-square test of contingency

Oftentimes we are interested in the interaction between factors. For example, in a study of dependence of people's weight on the area they live, we randomly select $N=1000$ from the nation and count the number of subjects from province A, who are underweight, normal or overweight. We repeat the categorization for each province and obtain Table 1.1.

Denote the nationwide overweight proportion as $p(o)$ and the proportion of province A residents as $p(A)$. If weight is independent of province, we would expect to find $N \times p(o) \times p(A)$ in cell (o, A) in Table 1.1. When the observed counts are far away from expectation, contingency of weight on area is implicated. The overweight probability $p(o)$ can be estimated by counting all the overweight people regardless of their residence. The sum is then divided by N, which should be the total counts in the table. The probability $p(A)$ is obtained similarly by dividing the sum of column A over different weight categories by the total number of counts N.

The statistical significance of the dependence is found by the χ^2-statistic,

$$\chi^2 = \sum_{i=1}^{r} \sum_{j=1}^{c} \frac{(O_{ij} - E_{ij})^2}{E_{ij}} , \qquad (1.21)$$

where the summation is over the total numbers of rows, r, and columns, c, in the contingency table and O_{ij} and E_{ij} are, respectively, the observed and expected counts in cell (i, j). If there is no dependence between the two factors in the population, the value of χ^2-statistic behaves like the Pearson's χ^2-distribution (after Karl Pearson) with $(r-1) \times (c-1)$ degrees of freedom as N is large. From the χ^2-distribution, we get the one-tailed p-value corresponding to the found χ^2 in the data and reject the null hypothesis of no dependence in the population if the p-value is smaller than the prespecified significance level α. A precaution is that for the test to be valid, the number of counts in each cell has to be larger than five. Otherwise, Fisher's exact test (after Sir Ronald Aylmer Fisher) has to be used.

1.2.9   Statistical power analysis

Specifying null hypotheses and seeking to reject the false ones are routinely undertaken in biomedical investigations. Power in statistics refers to the probability of rejecting a null hypothesis which is false. Power depends on four quantities, namely significance level α, effect size δ, variability σ in the population and sample size N, among which N is what we can plan. Sample size, in turn, translates to the scale of the experiment. In research proposals to funding agencies, we usually perform a power analysis to demonstrate or justify the budget we are asking for. If the power is not shown adequate, there is little argument for project funding.

The desired power is usually set at 0.8 or higher. We will discuss the required sample size at this power when we change one of the other quantities while keeping the rest fixed. Suppose we are testing the difference in the average blood pressures between males and females. If the blood pressures are truly different between sexes, the larger the difference δ (called effect size), the smaller the sample size we need before we convince ourselves of the difference. If the variability σ in blood pressures among individuals in the population is small, we will need a small sample size to observe a difference of δ between the the male and female populations. Setting a larger value of α means that we tend to consider the observed difference more of a true discovery than a fluke. A weaker level of significance, therefore, requires a smaller sample size.

Variability σ of the variable in a population is never known to us. It can nevertheless be estimated from the results of our or others' past related experiments. The significance level is usually set at $\alpha = 0.05$. We then plug into the power analysis formula a range of possible effect sizes to see the corresponding sample sizes. We then project the sample sizes that look feasible under the constraints of finite time span and manpower. Finally, we get an idea of the required sample size per group for 80 percent chances of detecting a larger than δ difference in a population of variability σ in the detecting variable at a significance level of 0.05.

Chapter 2

DNA Methylation Microarrays and Quality Control

A microarray experiment involves various steps, each of which is prone to error. For random error, repeated measurements will improve the precision. For systematic error, data normalization is to remove the bias. In either case, we need to have control of the quality of microarray measurements so that the results are reliable and credible. To this end, this chapter reviews the principles of gene expression microarrays which, after adaption, enable genome-wide DNA methylation profiling. We introduce two-color microarrays with probes from CpG (cytosine-guanine) island libraries and one-color oligonucleotide tiling arrays with probes covering nonrepetitive portions of the genome region of interest. Two popular enrichments of DNA methylation fragments for microarray hybridization are introduced; one is based on methylation sensitive restriction enzymes and the other on immunoprecipitation with antibodies against methylated DNA.

After hybridization, a microarray is scanned, resulting in an image of fluorescent intensities. An example of a measure for microarray quality is introduced. Array quality can be used to screen for outlier arrays that can be discarded for subsequent analysis. Meanwhile, visual inspection of an array image helps identify local peculiarity within the array.

If an experimental result is to be useful for the community at large, the experimental finding has to be reproducible by others. Although systematic studies of the reproducibility of microarrays for differential DNA methylation have not be done, identification of differentially expressed genes by high density, one-color, gene-centric microarrays was found to be reproducible between laboratories. Reproducibility and comparability of DNA methylation microarray experiments can be enhanced by control spots on the microarrays. Further improvement is expected by selecting the loci of large methylation fold changes between sample groups.

The situation is similar for the mapping of methylated DNA fragments using unbiased tiling arrays. In the mapping of RNA transcripts on the genome using tiling arrays, the unprocessed probe intensities from the tiling arrays were found to be highly correlated between biological replicates. The correlation of the hybridization intensities from enriched methylated or unmethylated DNA fragments may not be as high due to the variabilities in the source as well as in the enrichment.

2.1   DNA methylation microarrays

DNA methylation is one of the major epigenetic mechanisms that cause heritable alterations in various genomic functions, including gene expression, without a change in the coding and promoter sequences of the gene. Methylation can occur at the CpG dinucleotides in the promoter regions of mammalian genes. The affinity of the transcription factor binding sites changes, resulting in silencing of the gene's transcription. Stable heritable patterns of transcriptional silencing underlie normal development of multicellular organisms. Examples include genomic imprinting and X chromosome inactivation. Disrupted epigenetic profiles by gene mutations inflict human diseases, which can be innately inherited or somatically acquired. For example, several neurodevelopmental disorders (Beckwith–Wiedemann syndrome, Prader–Willi and Angelman syndrome, Rett syndrome) were found to originate from the DNA methylation and other epigenetic changes. Aberrant *de novo* methylation of tumor suppressor genes, occurring early in tumorigenesis, is a hallmark of human cancers. Autoimmune diseases and aging are among other examples. An understanding of the function and regulation of genome for better health thus will benefit most from epigenomic studies. Profiling of the methylation pattern along the complete genome is essential to elucidating the role of DNA methylation in organismal development and genome stability.

DNA microarrays are a high throughput technology, allowing for simultaneous measurements of the abundance of thousands of different DNA fragments in cells in a single assay [Schena95, Lockhart96]. Figure 2.1 illustrates the principles of DNA microarray. The basic idea is that the sequences of the fragments of interest are identified and immobilized on different locations on a solid (usually glass) surface. The affixed sequences are called probe sequences. The fragments from the sample are called targets, which are labeled with fluorophores before hybridizing to the microarray. During hybridization, the targets will preferentially bind to the spots whose probe sequences are complementary to the targets'. A readout system scans and records the fluorescent intensity from each probe on the microarray surface. The location and magnitude of the intensity tells what DNA fragments are present in what amount in the cellular sample.

Design of microarray probes for mammalian DNA methylation has focused on CpG islands [Cross94, Heisler05] or promoter regions where there exist over-than-expected numbers of CpG dinucleotides. Regions of dense CpG dinucleotides are called CpG islands (CGIs). Figure 2.2 maps the CGI probes of a human CGI microarray onto the human genome. The majority of CGIs co-localize with gene promoters. DNA methylation profiling thus has also utilized promoter microarrays. CGI-centric microarrays are biased in the sense that the methylation patterns in non-CGI regions are missed. An unbiased methylation profiling calls for tiling arrays whose probes cover the whole (non-

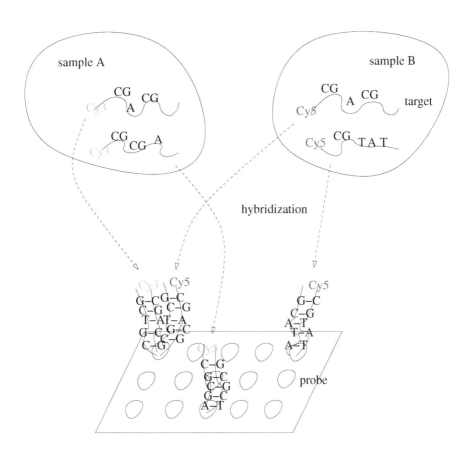

FIGURE 2.1: Principles of DNA microarray. DNA fragments from two samples are labeled with different dyes: Cy3 for green and Cy5 for red. In hybridization, target fragments CGACG from the samples will combine with the complementary GCTGC probe on the microarray due to Watson–Crick base pairing. Fluorescent intensities on the probes reflect the abundances of bound sequences originating from the samples. In the example, we conclude from the intensity data that CGACG fragments are of equal amount in the two samples, CGCGA exist only in sample A and that CGTAT exist only in sample B. Note that, in reality, target and probe sequences are longer and that nonspecific hybridization occurs so that abundance comparison or profiling are far from ideal.

FIGURE 2.2: Genomic positions of the probes on a human CpG island microarray. The start and end positions of the probe sequences on the human genome are connected by thick black lines. Note that only 9764 out of the 12,192 probes on the array had position annotations at the time of writing. And, 7843 out of the 9764 probes are unique and on the nonrepetitive portions of the human genome. The 7843 probes are in thin yellow.

repetitive) genome. Discussion of data from both types of microarrays appears in the book.

2.2   Workflow of methylome experiment

DNA microarrays were originally developed for high throughput gene expression measurement. The fragments in the last section would refer to the sequences reverse-transcribed from mRNAs in the context of transcriptome studies. To obtain genome-wide methylation profiles (so-called methylome), we rely on the protocols for enriching methylated DNA in the sample. Enrichment methods advance at a rapid pace. We introduce two of them that are mature in the techniques and versatile in the applications.

2.2.1   Restriction enzyme-based enrichment

The first is based on restriction enzymes that digest DNA sequences at specific sites into short fragments. The method is particularly suited for comparing methylation profiles in samples of different pathological or physiological conditions [Yan02, Schumacher06]. Genomic DNA from two samples are cut at non-CGI sites into fragments (100 to 200 bp). Some of the resulting fragments contain CpG islands, which are either methylated or not. The cut ends of the fragments are ligated with linkers. Methylation-sensitive restriction enzymes are then applied to chop up those fragments that contain unmethylated CpGs. The remaining fragments, mostly containing methylated CGIs, are amplified by polymerase chain reaction (PCR). After labeling reaction, the two differently labeled samples are co-hybridized to the CGI microarray. Data are analyzed to identify differentially methylated loci. Note that single nucleotide polymorphisms (SNPs) between two samples at the same cutting sites may yield differences in hybridization intensities, mimicking methylation differences. This confounding factor has to be addressed in the analysis. Finally, independent studies are run to, for example, confirm the regulatory role of the selected CGIs. Figure 2.3 depicts the workflow of a DNA methylation microarray experiment using methylation-sensitive restriction enzymes.

2.2.2   Immunoprecipitation-based enrichment

The second approach to enrichment is derived from chromatin immunoprecipitation, (ChIP) which is widely used in studies of protein–DNA interactions and histone modifications. The protein-bound DNA is randomly broken into small fragments. Fragments containing the proteins are pulled down by the antibodies specific to the protein in question. The isolated fraction is puri-

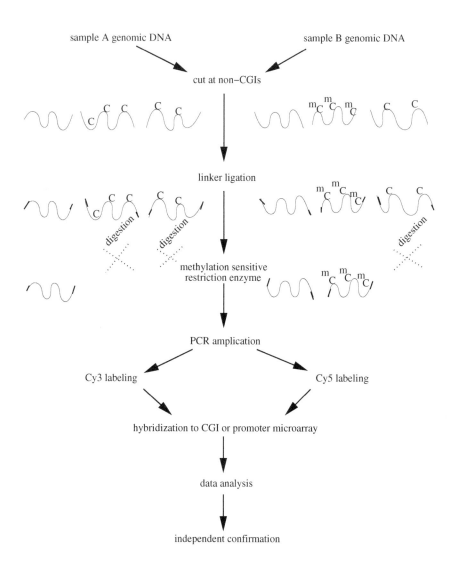

FIGURE 2.3: Methylation-sensitive restriction enzyme-based enrichment.

fied and hybridized to the microarray after labeling. When the ChIP assay is modified using antibodies against 5-methylcytosine, the methylated DNA immunoprecipitation (mDIP) method enables capturing of methylated DNA [Weber05]. mDIP, coupled with genomic tiling microarrays, is suitable for unbiased whole genome methylation profiling, uncovering non-CGI methylation regions. Figure 2.4 shows the key steps for the methylation enrichment using mDIP.

2.3   Image analysis

The microarrays are post-hybridization washed and dried and are ready for image acquisition. Often seen scanners are AXON GenePix® scanners ($>$ 4000 series) for dual-channel DNA microarrays and Affymetrix GeneChip® scanners ($>$ 3000 series) for single-channel oligonucleotide arrays. A microarray scanner consists of solid-state lasers for fluorophore excitation, photomultipliers for photon detection, optics for uniform detection across the surface, and the associated data acquisition electronics. Design goals of the scanner are to achieve high scan speed, high spatial resolution, high signal-to-noise ratio and high intensity dynamic range (sixteen-bit). Results of scanning are sixteen-bit grayscale image files in TIFF format per channel. The image files are to be stored for an indefinite time, whereas the duplexes on the hybridized arrays can decay quickly.

The image files are then analyzed by the software accompanying the scanner. The microarray image analysis software performs alignment of image blocks; locates the features (i.e., probes) within blocks; divides the pixels of a feature into signal (or foreground) and background; calculates the mean, median and standard deviation of the signal and background intensities; and outputs the calculations of each feature into a text file row by row. The output files (e.g., ∗.GPR from GenePix and ∗.CEL for Affymetrix), together with the TIFF files, are considered raw data of the microarray experiment.

Of interest for data quality control are the signal-to-noise ratio column and "flag" column in an output GPR file. GenePix uses a proprietary feature finding algorithm that defines circles, the pixels within which register light from specific hybridization. Light outside the circles are background intensities from nonspecific hybridization or from stray photons during scanning. The image analysis software then calculates the signal-to-noise ratio for each spot by dividing the mean signal minus mean background intensities by the standard deviation of the background pixel intensities. Each spot is also associated with a flag indicating whether its quality is "good" or "bad." By bad, it means the circularity of the signal by the feature-finding algorithm is bad or that the circle is so large that it overlaps with other circles.

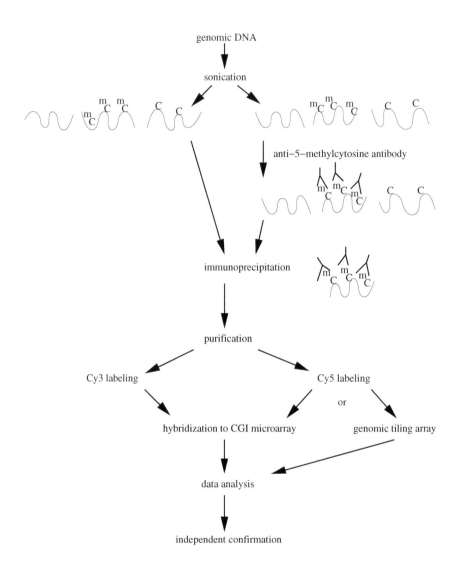

FIGURE 2.4: Methylcytosine immunoprecipitation-based strategy for DNA methylation profiling.

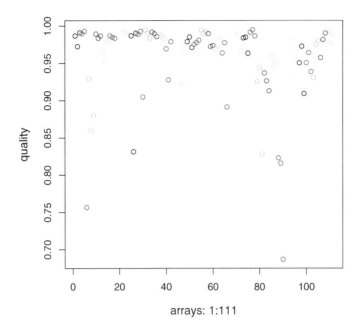

FIGURE 2.5: Array quality control. Array quality is defined as the percentage of unflagged spots on the array. The median quality of the total of 111 arrays is 0.98. There are, however, a few outlying arrays as seen by their low percentages (<95 percent).

An epigenetic profiling experiment can involve dozens of microarrays. It is essential to ensure that the quality of the assay is uniform between the microarrays within and between hybridizations. As a measure of quality, we count the number of spots that are flagged bad upon reading in a microarray. The ratio of the bad spots to the total number of spots on the microarray is defined as the quality of the microarray. A low quality may indicate imperfection in probe spotting or array hybridization. Figure 2.5 shows an example of the plot of quality versus arrays. We may exclude the bad spots or the whole low-quality microarrays for subsequent data analysis. Fortunately, as the coating and printing technologies improve, variations in probe fabrication are greatly reduced. For example, the Agilent's SurePrint® technology synthesizes 60-mer oligonucleotide probes *in situ* on the array using a noncontact inkjet approach, greatly enhancing spot uniformity.

2.4 Visualization of raw data

Inspection of microarrays by eye is indispensable to identify artifacts that escape image analysis software. For example, nonhomogeneous hybridization over the microarray surface can cause a changing shade from one edge (or corner) of the microarray to the opposite edge (corner). The image can also show scratches due to surface contamination. To visualize a microarray, the positions and mean intensities of the spots are read. Dots are painted on a two dimensional canvas with relative positions and greenness and redness corresponding to the input data. The plot includes both signal and background intensities of each probe. Figure 2.6 to Figure 2.8 show examples of the images from a two-channel human CpG island microarray. It is also desirable to juxtapose all the array images to ease comparison, as shown in Figure 2.9.

2.5 Reproducibility

An international project by the MicroArray Quality Control (MAQC) consortium, under the U.S. Food and Drug Administration (FDA) sponsorship, studied the intra- and interplatform reproducibility of genome-wide gene expression profiling by different microarray platforms [MAQC06]. The unprecedented large-scale study involved 137 laboratories using 1327 microarrays of 7 different platforms ranging from homebrew, two-color spotted microarrays to commercial one-color oligonucleotide arrays. The results from the pilot phase of the project indicated that, within laboratories using the same platform, the expression measurements, having a coefficient of variation of 5 to 10 percent in the expression signals from replicates of the same sample, are repeatable. Furthermore, between laboratories using the same platform, the identified differentially expressed genes, having an average concordance of 89 percent, are reproducible. Finally, between one-color platforms, the identified differentially expressed genes, having an average concordance of 74 percent, are comparable. The results are encouraging considering the diversity in the probe design strategies and in the protocols leading to hybridization across the platforms.

DNA methylation microarrays are, however, different from gene expression microarrays in their prelabeling steps toward hybridization. In particular, DNA methylation microarray experiments based on restriction enzymes involve the extra steps of digestion, ligation and PCR amplification (cf. Figure 2.3). In technical replicating DNA methylation microarray measurement with genomic DNA from the same biological source, digestion and ligation should in theory perform similarly each time, provided the conditions are correct

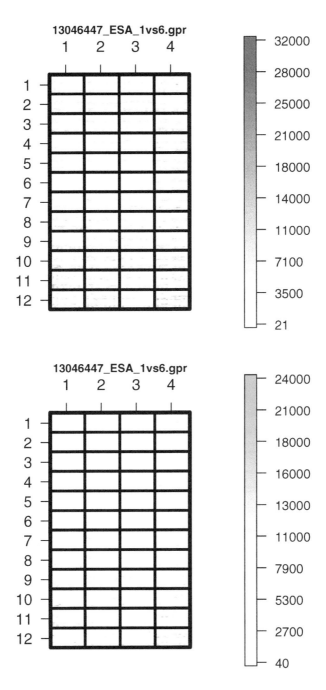

FIGURE 2.6: Foreground intensities of the Cy3 (top) and Cy5 (bottom) channels of a human CpG island 12k microarray.

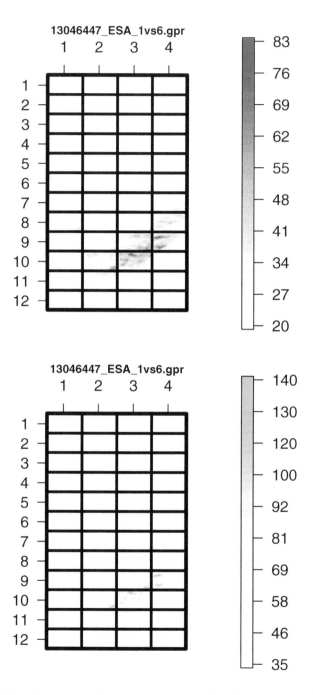

FIGURE 2.7: Background intensities of the Cy3 (top) and Cy5 (bottom) channels of the human CpG island 12k microarray of Figure 2.6.

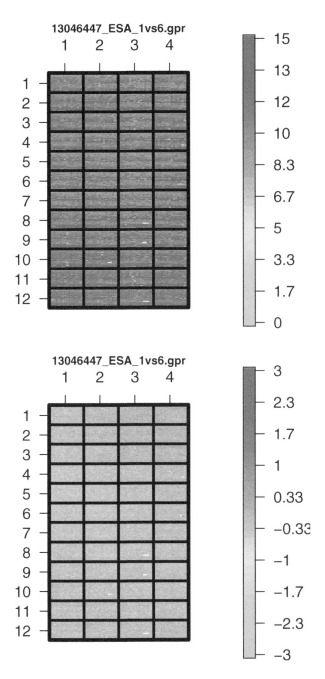

FIGURE 2.8: $\left[\log_2(\mathrm{Cy3}) + \log_2(\mathrm{Cy5})\right]/2$ (top) and $\log_2(\mathrm{Cy3}) - \log_2(\mathrm{Cy5})$ (bottom) of the background-subtracted intensities of Figure 2.6 and Figure 2.7.

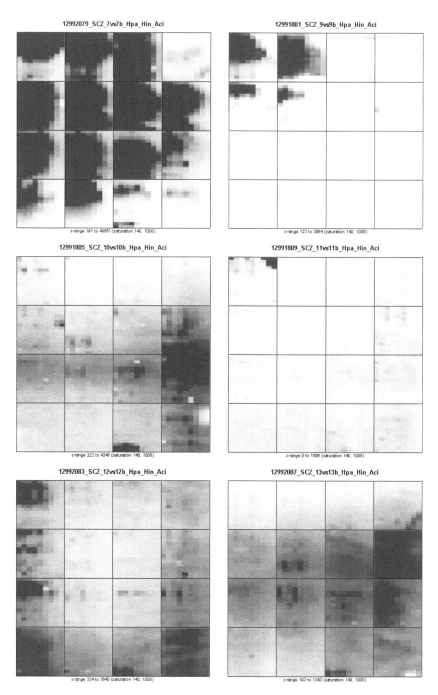

FIGURE 2.9:   Juxtaposition of multiple arrays in an experiment for easy comparison.

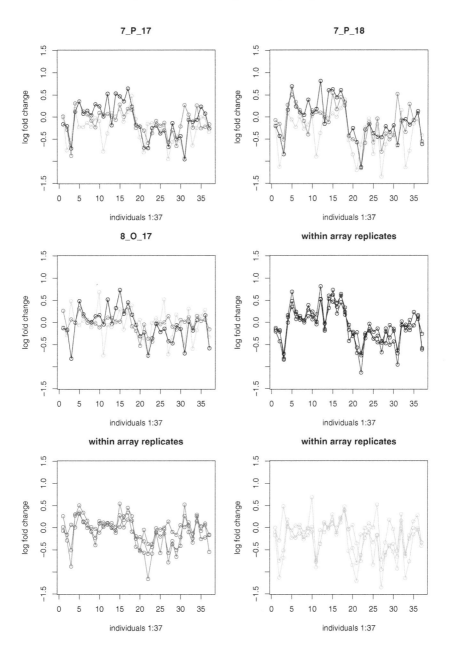

FIGURE 2.10: Microarray replication. A data point resulted from a probe on a microarray. Colors refer to batches of replications. Methylation fold changes of the *y*-axis are relative to a common reference sample. *x*-axis is across different individuals. Probes 7_P_17, 7_P_18 and 8_O_17 have identical position annotation, representing intraarray technical replication.

(e.g., with enough enzymes and adaptors). The fragments bound by the primers in the initial stages of the PCR have higher chances of being amplified. While much of this is the same, slight changes in the early cycles between PCRs can produce different amounts of enrichment fractions. The PCR step is thought to contribute most of the variability in enrichment based on restriction enzymes. In Figure 2.10, we show results of replication at different levels: intraarray technical replicates, interarray technical replicates and interarray biological replicates. The concordance decreases as the level of replication drops from within arrays to between samples as we expected. The median Pearson correlation of the methylation (raw log intensity ratios) between interarray technical replicates by human CpG island microarrays over thirty-seven independent individuals was found to be 0.76. To reduce the variability, for a sample, we may have, say, two technical replicates up to the PCR step. The two batches of amplicons then are pooled into one, before hybridizing to a microarray.

2.5.1   Positive and negative controls by exogenous sequences

In the wake of the potentially larger variability in DNA methylation compared with gene expression microarray experiments, manufacturers of CpG island microarrays have included control spots in their design. That is, in addition to the CpG island clones, PCR products of artificial sequences are spotted on the microarray. The artificial sequences are made by screening against existing annotated sequence databases to make sure they are not homologous to any eukaryotic and prokaryotic genomes known to date. Preset amounts and ratios of the sequences corresponding to the artificial probes are then spiked into the case and control samples. The different amounts and ratios define the red and green intensities and the intensity ratios, serving as positive controls. Allowance is also made to spot on the microarray negative controls using DNA of foreign species, such as salmon sperm. The control probes together with the spike-ins are useful for between-experiment normalization and comparison, improving reproducibility.

2.5.2   Intensity fold-change and p-value

The MAQC result also revealed an increase in cross-platform concordance if the lists of differentially expressed genes were derived from genes ranked according to their fold-changes, besides their p-values [MAQC06]. The implication for the restriction enzyme-based enrichment is that different enzymes can be exploited for mapping of methylated or unmethylated genome. Suppose 10 percent of the CpG dinucleotides in a CpG island is methylated in control samples. A 10 percent increase in the methylation level in cases results in a two-fold change (from 10 percent to 20 percent) in the methylation ratio. However, at a CGI locus whose methylation level is already as high as, say, 80 percent, a 10 percent change from 80 percent to 90 percent gives rise to only a fold change of 1.125 between the controls and cases. The enrichment strategy of Figure 2.3 is, therefore, optimal in targeting the unmethylated regions of genome for hypermethylation detection and is suitable in such applications as abnormal silencing of tumor suppressor genes and cancer-specific hypermethylation in cancer research.

2.5.3   DNA unmethylation profiling

On the other hand, since the majority of the cytosines in human genome are methylated except for the tissue-specific and housekeeping genes, it is arguably more informative to target the methylated genome for hypomethylation detection in exploratory studies. Now, again, suppose the methylation status at a locus changes from 90 percent to 80 percent between groups. If what we measure is the degree of unmethylation, the change corresponds to a two-fold change (from 10 percent to 20 percent) in intensities. The restriction enzyme-based enrichment proves flexible in this regard. To measure degrees of unmethylation [Schumacher06], genomic DNA are digested with methylation sensitive restriction enzymes such as HpaII, which cuts at unmethylated CCGG. The digested fragments are ligated to DNA adaptors. PCR, with primers complementary to the adaptor sequences, then preferentially amplifies the shorter (50 b to 1.5 kb) fragments, i.e., the unmethylated DNA fragments. The remaining steps, i.e., labeling and hybridization, follow those in Figure 2.3.

2.5.4   Correlation of intensities between tiling arrays

In applications where we are identifying genomic regions of DNA methylation in an unbiased fashion, enrichment of the regions by methyl DNA immunoprecipitation followed by genomic mapping by high-density tiling arrays does the job. A related task is the mapping of the transcribed regions of the genome by high density tiling arrays. An assessment of the performance of such mapping using Affymetrix and NimbleGen® tiling arrays reported that the average Pearson correlations of the unprocessed hybridization intensities between arrays (either technical or biological replicates) within platforms were as high as 0.96 for Affymetrix and 0.83 for NimbleGen [Emanuelsson06]. A high correlation in transcript mapping indicates high reproducibility in hybridization. However, in methylation mapping, variability in biological sources (cells in a tissue do not share a common epigenome) and noise from the mDIP enrichment are expected to compromise the reproducibility.

In terms of detecting a methylated fragment, the higher the density of the tiling array, i.e., the more probes for, say, every 100 bp of the genome, the higher the sensitivity of the detection. The reason is that for a methylated fragment, more measurements are obtained from the larger number of probes, manifesting the sample size problem of section 1.2.9.

Chapter 3

Experimental Design

Microarrays are a high-throughput assay technology that measures the expression levels of thousands of genes in a single experiment. On the other front is the increasing number of species whose genomes are decoded, thanks to the continuously decreasing cost of DNA sequencing. The two technologies together are revolutionizing biological experimentation. A microarray measurement involves a series of steps including DNA/RNA extraction, enzyme digestion, labeling, hybridization, washing, drying and scanning. Skills and experience are indispensable in each step for high quality and reproducible results. Costs of samples, microarrays, chemicals and labor are not insignificant. A better design generates a dataset with less noise, giving rise to a result of higher statistical significance. A particular design can also limit the analysis methods that can be applied to the data. Careful planning of microarray experiments, therefore, can never be overemphasized.

Differently labeled DNA fragments compete to bind to the complementary sequences of the probes on a two-color microarray. The dual channels open up opportunities for different designs of microarray experiments. A design may lead to a better estimation of the parameters of interest than others. The choice thus depends on the experiment's specific aims. We start with three general designs, namely, reference design, balanced block design and loop design, that compare DNA methylation (or gene expression) between classes. The design involving single-color oligonucleotide chips is equivalent to a reference design with two-color DNA microarrays. We also describe design guidelines for factorial and time-course experiments that cater for particular methylation contrasts of interest.

The number of samples/microarrays required for confident hypothesis testing is central to any grant application as it translates to the amount of funds requested. We address the sample size issue based on sound statistics. Pooling of samples saves microarrays. However, we argue against pooling unless there are no other options. Note that, in the following when there is no confusion, the terms DNA methylation and gene expression are used interchangeably, so are the corresponding technologies.

3.1   Goals of experiment

3.1.1   Class comparison and class prediction

Designs of microarray experiments will depend on experimental objectives. Microarrays are used to look for sequence fragments (loci) that are differentially methylated across conditions. For example, genomic DNA samples from diseased cell populations are isolated and labeled with red dye (Cy5), and those from healthy cells with green dye (Cy3) in a two-color CpG island microarray experiment. Another example is to label samples on drug treatment with one of the dyes and without treatment with the other. In one-color oligonucleotide chip experiment, half of the chips are hybridized with diseased (treated) samples and the other half with healthy (untreated) samples. We call this kind of experiments *class comparison* because classes, such as diseased, normal, treated and untreated, are known and defined before experiment. After the differentially methylated loci (expressed genes) are identified by microarray experiments, they are verified by independent quantification techniques, such as bisulfite sequencing. The methylation profile of the differentially methylated loci can then be used as a signature in later microarray measurements for the purpose of diagnosis and/or prognosis. This type of applications can be called *class prediction*.

3.1.2   Class discovery

Suppose we obtain one hundred DNA methylation profiles for one hundred different phenotypes, one profile for each phenotype. We define a distance metric in the space of epigenomic dimensions. We then calculate the distance between any pair of the profiles. Because of variation in the samples, pairs whose distances are below a margin set by the variation are considered to belong to a taxonomy or cluster. If the one hundred samples are from one hundred breast cancer patients, we could be discovering breast cancer subtypes if clusters are found among the one hundred methylation profiles. Alternatively, if we have twenty samples from twenty normal individuals, we calculate the epigenetic distances between any two genes in the twenty-dimensional space. We again try to find clusters. Genes that fall into the same cluster are likely to share a cellular function. Clustering in this way hopes to yield clues on the function of unknown genes. Experiments of this kind are called *class discovery* and microarrays are a promising tool in this application.

Microarrays are also employed to study the effect of treatment on different strains of organisms. Yet, in time-course experiments using microarrays, researchers might be interested in the precise instance of time when the loci change the methylation or expression. As the technology advances, we expect to see more novel applications of microarrays in biomedical research. There is rarely an optimal design that is universal to all experimental goals. Different experimental aims dictate different designs. We, therefore, will summarize guidelines that address statistical issues in the design of microarray experiments.

What is specifically meant by "design" in a microarray experiment? An investigator hardly has unlimited time and resources to conduct an experiment. Either the amount of biological specimens is limited because they derive from a rare species. Or the total number of microarrays available is fixed because of lean budget. More often, both the specimens and microarrays are constrained. Even though neither is a major concern at the time of experiment, the investigator may want to perform it at a minimal cost in the hope that more follow-up experiments can be planned in the long run. Questions arise as to how to assign samples to microarrays and, on a microarray, how to assign labels to samples. For single-color oligonucleotide chips, the issue is greatly simplified. In fact, the design and analysis of single-color oligonucleotide chips are reduced to those of reference design of two-color microarrays.

3.2 Reference design

Reference design is one of the most common designs in microarray experiments. In this design, multiple aliquots of a single sample are prepared and labeled with one of the dyes, say, Cy3. An aliquot of sample i from individual i is also prepared and labeled with Cy5, where $i = 1, 2, 3, \cdots$, up to the number of samples in the experiment. Each Cy3-labeled aliquot is then mixed with a Cy5-labeled aliquot for co-hybridization to a microarray. The design does not necessarily mean the number of independent samples is equal to that of microarrays. For example, we can have an aliquot of sample 1 on microarray 1 and the other aliquot of sample 1 on microarray 2. The number of distinct biological samples thus is less than or equal to the number of microarrays in a reference design. Microarrays 1 and 2 are technical replicates. Figure 3.1 shows a reference design. In the figure, a box represents an aliquot of sample and an arrow indicates a microarray. The head and tail of the arrow refer respectively to Cy5 and Cy3 labeling. Since the Cy3-labeled sample is from a common biological source and present in every microarray, it is called reference sample. The Cy5-labeled samples are called nonreference samples. In a class comparison experiment involving, say, twenty-four patients, the reference sample can be prepared by pooling the genomic DNA (or RNA) from twenty-four matched healthy individuals for DNA methylation (or gene expression). Aliquots of the pooled DNA (RNA) are then labeled Cy3 for microarray hybridization.

The spots on a microarray slide can vary in size and quality of printing.

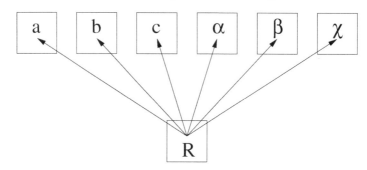

FIGURE 3.1:   Reference design. Samples a, b, c are from one class while samples α, β and χ are from the other class. R is a common reference. A microarray is conveniently represented by an arrow, which points toward a Cy5-labeled sample from a Cy3-labeled sample. The reference design is readily extended to the task of comparing more than two classes. The nonreference samples can be a, b, α, β, 1 and 2, belonging to three classes, for example. The chapter appendix includes more examples.

Furthermore, the concentration of target sequences on the surface of the slide can be nonuniform during hybridization. The resulting variability in DNA methylation measurement is hopefully canceled in two-color microarrays because whenever a spot registers a brighter red signal due to the nonuniformity, it might register a brighter green, too (cf. Figure 2.6 and Figure 2.7). The ratio (or log ratio) is not affected by the locus-by-spot effect. (It is called gene-by-spot effect in gene expression microarrays.) Two-color microarrays in a reference design as in Figure 3.1 are efficient in finding loci that are differentially methylated between conditions. Efficiency here is defined as the inverse of the variance of estimation. That is, the higher the efficiency, the more precise the parameters are estimated.

On the other hand, the experimental objective can be to find differences among individuals (usually of the same class, as in the case of subtype discovering by clustering). Note that objectives can also be multiple with a primary goal and many secondaries. Reference design is also appropriate in such applications, the reason being that the same reference sample is used in all microarrays so that it serves as a common reference for comparing the nonreference samples on the microarrays. This feature allows us to perform a class discovery analysis on data from reference design. The suitability of a reference design for the dual purposes of class comparison and class discovery lends its prevalence in microarray experiments. In fact, most commercial software for two-color microarray data analysis implicitly assumes that measurements are done in a reference design so that conventional statistical toolkits, such as t-test and clustering (cf. chapter 5, chapter 7), such as K-means can be readily applied.

3.2.1 Dye swaps

It is known that Cy3 fluorophore binds better to nucleotides than Cy5 does, resulting in a brighter Cy3 intensity even though the competing sequences in the mixture are of equal amount. This dye bias can be corrected for by normalization introduced in the next chapter. It, however, was also found that the bias could depend on target sequences. This sequence-specific dye bias cannot be removed by the normalization. We can observe the sequence-specific dye bias by plotting the distribution of log ratios from an experiment where two aliquots of the same sample are individually labeled to Cy3 and Cy5. We expect around zero log ratios (i.e., no fold change) for most of the sequence fragments from such a self hybridization. Any fragment with log ratio far away from zero is a potential victim of dye bias and should be marked with caution in subsequent analysis.

When we compare among nonreference samples in a reference design, the sequence-specific dye bias effect does not matter since all the nonreference samples are labeled with the same dye. If, however, we are comparing nonreference with the reference sample, the sequence-specific dye bias becomes trouble. To eliminate the artifact, we allocate, say, two pairs of microarrays for replication. The microarrays in each pair are technical duplicates of each other except that the dye assignment in one microarray is reversed to that in the other microarray in the same pair. The two pairs of dye-balanced microarrays can be used to assess the sequence-specific dye bias effect (cf. section 5.2.2). The result is then used for the correction on all the microarrays. Note that given a total of, say, twenty microarrays in this case, we choose to form two dye swapping pairs. The number of independent biological samples is eighteen (= sixteen singlets + two duplicates). If we form ten pairs of dye swaps, the number of independent biological samples reduces to ten. The power of statistical inference suffers. This is because the biological variability (intersample standard deviation of methylation or expression) is, in general, larger than the technical variability (intrasample standard deviation of methylation or expression) in microarray experiments. To lower the total variability for increased power, the priority is to reduce the biological variability by increasing the number of biological samples. Technical replicates do not help much and two or three pairs of dye swaps are considered enough to estimate sequence-specific dye bias.

3.3 Balanced block design

Block design has a longer history in statistical experimental design than microarray technology. Since DNA microarrays use two colors, we can compare two classes of samples arranged in the so-called balanced block design shown

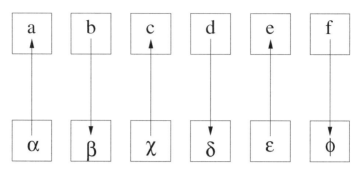

FIGURE 3.2: Balanced block design. Samples a, b, c, d, e, f are from one class while samples α, β, χ, δ, ϵ and ϕ are from the other class.

in Figure 3.2. In the design, samples 1 of class 1 and 2 are labeled respectively red and green on microarray 1; samples 2 of class 2 and 1 are labeled respectively red and green on microarray 2, $\cdots$, and so on. Labeling half the samples with red dye and the other half with green dye for each class is a way to eliminate the sequence-specific dye bias. The balanced block design example of Figure 3.2 takes this approach to make it immune to the bias.

The advantage of a balanced block design is that, given the same number of microarrays, the design has a higher power of distinguishing classes than a reference design [Simon03]. This is because the comparison of classes in this design is direct, whereas the comparison among the nonreference samples via the reference sample in a reference design (cf. Figure 3.1) is indirect. The advantage degrades as the number of classes to compare increases. In particular, if the number of samples is the same, the power of a balanced block design becomes similar to or lower than that of a reference design as the number of classes to compare goes beyond two. This is because in such cases the number of microarrays used by a balanced block design is less than that by a reference design.

The major drawback of a balanced block design is that data from such a design are not appropriate for class discovery. This is because all pairwise distances are required for cluster analysis. In a balanced block design, however, not all such distances are calculable. Whereas in a reference design, distances between any pairs of nonreference samples can be obtained through the reference sample.

It can happen that the assignment of classes to samples had erred and the mistake was not realized until after data-taking. When the correction is made during data analysis, the configuration can no longer be a balanced block design, resulting in a loss of efficiency for class comparison. The worst scenario is that samples of the same class are comparing to each other such that class comparisons cannot be made. Similarly, if one or two microarrays failed dur-

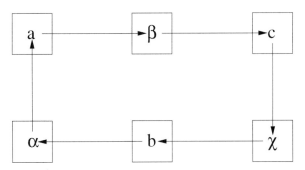

FIGURE 3.3: Loop design. Samples a, b, c are from one class while samples α, β and χ are from the other class for class comparison.

ing hybridization, the remaining microarrays cease to form a balanced block design, compromising its efficiency. A reference design remains a reference design in such cases. Balanced block design thus is not as robust as reference design.

3.4 Loop design

Loop design was proposed by Kerr and Churchill in an effort to minimize the average of the between-sample estimation variances (i.e., the so-called A-optimality) [Kerr01a]. The design, therefore, is supposed to be optimal when comparisons between all possible pairs of samples are of interest. In the design, Figure 3.3, an aliquot of a sample is labeled red in one microarray and the other aliquot of the same sample is labeled green in the other microarray. Using the same sample on the two microarrays as an intermediary, the two microarrays are considered to be connected. If we continue the pattern with new microarrays until the last microarray, which hosts the last sample and the first sample, the connecting pattern is closed, forming a loop.

The power of a loop design is always higher than that of a reference design given the same number of microarrays or samples for class comparison. Specifically, when the number of microarrays is fixed, the efficiency of loop design is higher than reference design, but lower than balanced block design. The advantage, however, fades as the number of classes increases or as the ratio of biological variability to technical variability goes beyond two. As in balanced block design, sequence-specific dye bias effect can be estimated and adjusted in loop design because classes are balanced with respect to dye.

Unlike balanced block design, loop design generates data from which distances between any two samples can be calculated, through the intermediary samples. Therefore, loop design is effective in class discovery. However, the precision of distance estimation strongly depends on the relative position of the samples in the loop. The farther apart the two samples are along the chain, the larger uncertainty in the distance estimation. This is because the estimation will have to go over the intermediary microarrays between the two samples. The more intermediary microarrays there are, the more noise can accumulate from the individual hybridizations.

Loop design suffers the same weakness in robustness as balanced block design. If class assignments have to be changed or few outlier microarrays have to be discarded, the loop can be broken, paralyzing the analysis.

3.5   Factorial design

We have so far focused our discussion of microarray experimental design on class comparison and discovery. In some cases, the experimental objectives can be more refined. For example, we may be interested in the different drug responses across cell lines. The primary objective is then the "interaction" between drug and phenotype. Differences in the methylation (or expression) between drugs and between phenotypes are secondary objectives. This is an example of factorial design where multiple factors, each of which can have many different levels, are used to explain experimental observations.

For the purpose of illustration, we first consider a 2×2 factorial design. Suppose an experiment is planned to detect difference in DNA methylation between patients and controls. Furthermore, the difference can depend on sex. We have two factors: disease status and sex. Each has two levels: patient versus control and male versus female. We acquire DNA samples from male patients (mP), female patients (fP), male controls (mC) and female controls (fC). From the four types of samples, six different pairings can be formed for hybridization on a two-color microarray. There are actually twelve if we count the dye-swap of each pair. Suppose the number of microarrays we can afford is fixed in this experiment. The question is then which pairs should we choose in order to minimize the uncertainty in the estimation of differential methylation. In other words, how should we arrange the samples on the microarrays in order to maximize efficiency.

To study the design, we employ a so-called linear additive model of DNA methylation (or gene expression). The model is simple, yet it captures the essence of many complex biological phenomena. First of all, we use logarithm of fluorescence intensity as the measure of DNA methylation (or DNA unmethylation, gene expression) level. (The intensity is already background-corrected and normalized (cf. chapter 4).) This convention is found em-

TABLE 3.1:   Linear Model for DNA Methylation or Gene Expression

Experimental Conditions in the 2×2 Factorial Design	Log-intensity
mC	μ
mP	$\mu + \alpha$
fC	$\mu + \beta$
fP	$\mu + \alpha + \beta + I_{\alpha\beta}$

pirically adequate. Especially, the transformation turns many multiplicative effects encountered in biology into additive ones. The log intensity of a locus is then described by a sum of parameters or coefficients that represent the effects of the factors as in Table 3.1. In the table, parameter μ represents the methylation level of mC. If we take mC as a reference point, then μ represents the baseline level of DNA methylation. α represents the effect of disease on male patients. β represents the difference in methylation levels between female and male controls. $I_{\alpha\beta}$ represents the sex-specific effect on the disease. Note the model applies to every probe on the microarray. That is, every DNA fragment has its own μ, α, β and $I_{\alpha\beta}$. Note also that since we can have many DNA samples from different individuals of each condition for the experiment, data from each individual are applied to the same model. The parameters estimated, therefore, represent mean effects over the individuals of the same class.

From the parameterization of the model in Table 3.1, it is clear that the effects of disease on male and female are picked up, respectively, by α and $\alpha + I_{\alpha\beta}$. If no sequences with significant nonvanishing $I_{\alpha\beta}$ are found, then the disease affects evenly between sexes. In this case, those sequences whose αs are significantly different from zero confer susceptibility to the disease. If the primary objective of the experiment is to find the disease effect, we should direct our limited resources (i.e., microarrays) toward measuring α and $I_{\alpha\beta}$, at the expense of measuring β.

DNA microarrays utilize two labels. Differently labeled target sequences from two samples (e.g., mC and mP) compete to hybridize to the complementary probe sequences on the microarray. Table 3.2 lists all possible pairs of samples on a microarray together with the parameters each measures from the log intensity ratio.

With Table 3.2, it becomes clear that our top priority is to invest in configurations 1 and 4, which measure α and $\alpha + I_{\alpha\beta}$. The next priority goes to the rest of the configurations, which, in addition, measure the less interesting β. If we have a total of six microarrays available for the experiment, we see that we would choose the design on the top of Figure 3.4 rather than the reference design in the middle of Figure 3.4. The optimal design would change as the objective of the experiment shifts. For example, suppose now estimation of α, β and the interaction effect $I_{\alpha\beta}$ are of equal importance. The optimal design becomes the bottom of Figure 3.4.

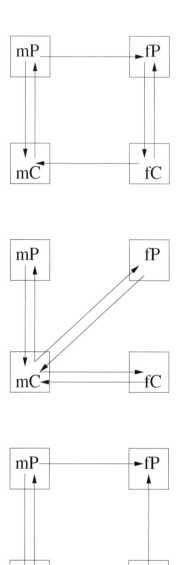

FIGURE 3.4:   Factorial design with emphasis on different parameters. (See text for details.)

TABLE 3.2:   Parameters Measured by a Microarray Using $\log(R/G) = \log(Cy5/Cy3) = \log(Cy5) - \log(Cy3)$ in a 2×2 Factorial Design

Configuration	Cy5 ← Cy3	Log-Intensity-Ratio
1	mP ← mC	α
2	fC ← mC	β
3	fP ← mC	$\alpha + \beta + I_{\alpha\beta}$
4	fP ← fC	$\alpha + I_{\alpha\beta}$
5	fP ← mP	$\beta + I_{\alpha\beta}$
6	fC ← mP	$\beta - \alpha$

Note: cf. Table 3.1.

Remember to guard against sequence-specific dye bias, arrangement of samples should be symmetric with regard to the dye. The objective can become so demanding that, in a factorial design, we also desire the robustness and/or class discovery feature of a reference design. Figure 3.5 shows a design that is a blend of reference and factorial designs. In the design, individual male (female) patients are hybridized to a pool of male (female) controls. This can be seen as separate measurements on α and $\alpha + I_{\alpha\beta}$. If we can combine the estimate for α from each experiment in one analysis, we can improve the precision of the estimation. We accomplish this by establishing links between the male and female pools as shown in the figure. The example also demonstrates the extensibility of reference design. The experiment involving males in the upper half of Figure 3.5 could have been done some time before the experiment involving females in the lower half of the figure. If the male reference sample has been well preserved since, the two experiments can then be integrated through the microarrays measuring the male versus the female reference.

Design considerations for the 2×2 factorial case can also be straightforwardly extended to designs with more than two factors such as $2 \times 2 \times 2$ and more levels per factor such as 2×3. For example, Table 3.3 shows the linear model for a 2×3 factorial design where the two factors are disease and regimen. The symptoms of the disease can be mild (C), moderate (M) and severe (S) and the types of regimen can be regular (1) versus special diet (2). If we are interested in the methylation changes between the regimen at different phases of disease progression, i.e., the interaction terms $I_{\alpha\gamma}$ and $I_{\beta\gamma}$ in the linear model of Table 3.3, an optimal design is shown in Figure 3.6 where we use only ten microarrays.

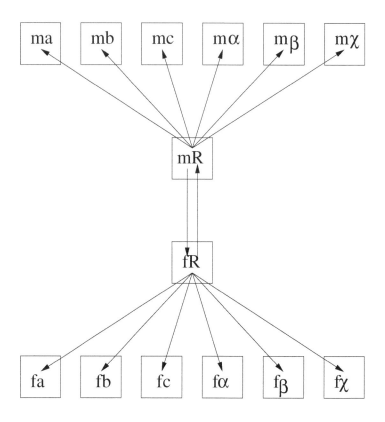

FIGURE 3.5: An extended reference design; a, b, c can be three patients in one group and α, β and χ can be three controls in the other group. R is a reference sample. The prefix m refers to male and f to female.

TABLE 3.3: Linear Model for DNA Methylation or Gene Expression

Experimental Conditions in the 2×3 Factorial Design	Log-Intensity
1C	μ
1M	$\mu + \alpha$
1S	$\mu + \beta$
2C	$\mu + \gamma$
2M	$\mu + \gamma + \alpha + I_{\alpha\gamma}$
2S	$\mu + \gamma + \beta + I_{\beta\gamma}$

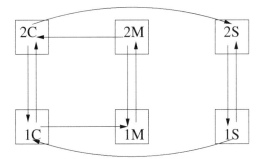

FIGURE 3.6: A 2 × 3 factorial design. 1 and 2 can refer to different diets or treatments. C, M, S can correspond to three different phenotypes, cell lines or time points.

3.6 Time course experimental design

Time course experiments are useful to study the roles genes play in a molecular pathway. In fact, one of the early demonstrations of microarray technology is a time course experiment profiling yeast genome expression across time points during glycolysis [DeRisi97]. Evidence suggests that the methylation profiles between monozygotic twins diverge with age. DNA methylation patterns are not as ephemeral as gene expression. Nevertheless, time course measurements on DNA methylation are not unimaginable.

Consider a time course experiment where RNA are extracted from cells that are harvested at four time points. There are six ways to pair the time points for co-hybridization. Similar to Table 3.1 and Table 3.2 for a factorial design, we parameterize the effect of time lags on gene expression with a linear model. The expected log intensity ratios from all possible different hybridization configurations can be shown in Table 3.4, where α_i represents the expression changes between time i, ti, and time zero, $t0$, for $i = 1, 2, 3$.

Now, if the primary interest is in the effect of time on the expression relative to the expression at time zero (i.e., $\alpha_1, \alpha_2, \alpha_3$), we would allocate more microarrays to configurations 1, 2 and 3. If, instead, the interest is in the effect relative to the previous time point (i.e., $\alpha_1, \alpha_2 - \alpha_1, \alpha_3 - \alpha_2$), we would focus on configurations 1, 4 and 6. Given that only six arrays are available for the time course experiment, we would choose the left design of Figure 3.7 for the former objective and the center one for the latter. Owing to the symmetry, the design in the right of Figure 3.7 can be shown to be efficient for both objectives [Glonek04].

TABLE 3.4:   Parameters Measured by a Microarray Using
$\log(R/G) = \log(Cy5/Cy3) = \log(Cy5) - \log(Cy3)$ in a Time
Course Experiment

Configuration	Cy5 ← Cy3	Log-Intensity-Ratio
1	t1 ← t0	α_1
2	t2 ← t0	α_2
3	t3 ← t0	α_3
4	t2 ← t1	$\alpha_2 - \alpha_1$
5	t3 ← t1	$\alpha_3 - \alpha_1$
6	t3 ← t2	$\alpha_3 - \alpha_2$

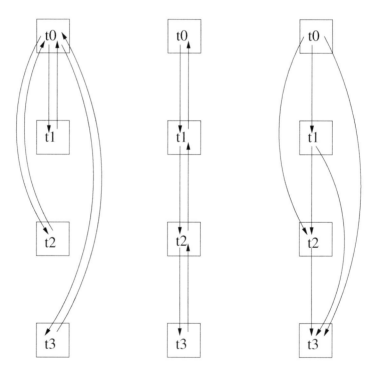

FIGURE 3.7:   Time course experimental designs that optimize estimation
of DNA methylation or gene expression differences between absolute time
(left), relative time (center) or both (right).

3.7   How many samples/arrays are needed?

3.7.1   Biological versus technical replicates

Degrees of DNA methylation exhibit variation across distinct samples of the same class (e.g., tissues or cells from different tissues and/or different individuals). The variation is of biological nature and is called biological variability. When a microarray measurement is repeated with samples derived from the same biological source, the measured intensities may vary from microarray to microarray. The variation is of technical origin and is called technical variability. Variation in the result of a microarray experiment comes from both biological and technical variations.

Recall that the error in sample mean estimation is related to the sample variance, which decreases with sample size. Because of biological variability, in order to make statistical inference about population, we need to sample as many independent biological sources from the population as possible. Furthermore, in most two-color DNA microarray experiments, biological variability overwhelms technical variability, the ratio of biological to technical variability ranging from two to eight. (Biological variability in inbred animals or monozygotic twins can be expected to be smaller.) Figure 3.8 shows examples of distributions of variability in log intensity ratios from methylation measurements on some human tissues. From the figure we can estimate the biological variability, σ_{bio}, by $\sigma_{\text{bio}}^2 = \sigma_{\text{tot}}^2 - \sigma_{\text{tech}}^2$, where σ_{tot} and σ_{tech} are respectively the total and technical variability. The examples show that the biological variability is about 4.5 times the technical variability. Therefore, it is understood that we strive to obtain as many genomic DNA samples from independent biological sources as possible for a microarray experiment. Technical replicates on the same biological samples tell little about population/idiosyncrasies. In a microarray experiment, two or three pairs of dye-swapping technical replicates that purport dye bias correction usually suffice.

3.7.2   Statistical power analysis

So, how many samples should be prepared for an experiment? To better answer the sample size question, we formulate the problem into a power analysis. Stories unfold as researchers are curious about (or upset with) a null hypothesis of no differences in means between two classes. Samples are then collected in order to prove (or disprove) the null hypothesis. Because of variation in individuals, a wrong conclusion about a difference can be made when there is, in fact, no difference, committing a false positive. Researchers generally tolerate no more than 5 percent cases of false positives. The 5 percent is called false positive rate, α. On the other hand, because of insufficient samples, an experiment can fail to find a difference when there is in fact one,

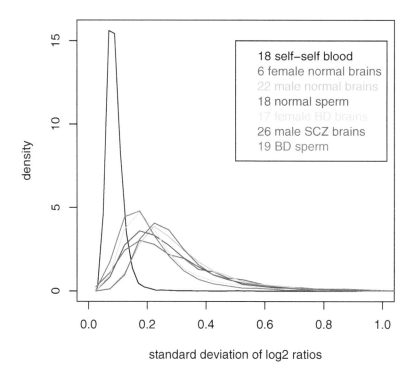

FIGURE 3.8:   Frequency distributions of variability in log intensity ratios from two-color CpG island microarrays with reference designs. The self hybridization measured technical variability while the rest registered total variability. A distribution results from 7843 standard deviations from the 7843 unique sequence probes on the microarray. Methylation data are background-corrected and normalized. Note the distribution can depend on experimental protocols, such as hybridization temperatures.

making a false negative. Let δ denote the size of difference between the means. (It is also called effect size.) The chance of committing false negatives is false negative rate, β. Statistical power, $1 - \beta$, is the probability of demonstrating a true mean difference of size $\geq \delta$. We usually set the false negative rate at 0.1 so that the power, i.e. chance of correctly rejecting the null hypothesis, is 90 percent. Let σ represent the standard deviation of the sample measurements, which are assumed normally distributed. Statistical power, $1 - \beta$, is a function of α, δ, σ and sample size n per class. n is the number of samples per class (or per group). In fact, given four of the five parameters, the rest can be calculated. Specifically,

$$ n = 2 \left[\frac{z_{1-\alpha/2} + z_{1-\beta}}{\delta} \right]^2 \sigma^2 , \qquad (3.1) $$

where $z_{1-\alpha/2}$ indicates the $100(1 - \alpha/2)^{\text{th}}$ percentile of the t-distribution (or standardized normal distribution for large n) [Dobbin05].

Equation (3.1) applies readily to reference designs of two-color microarrays as well as one-color oligonucleotide chips. In the case of a two-color microarray experiment, σ^2 is the variance of the log-intensity-ratios, while in the case of an oligonucleotide chip experiment, it is the variance of the log-intensities. Note that both intensities are background-corrected and normalized. σ includes contributions from both biological variability σ_{bio} and technical variability σ_{tech},

$$ \sigma^2 = \sigma_{\text{tot}}^2 = \sigma_{\text{bio}}^2 + \sigma_{\text{tech}}^2 . \qquad (3.2) $$

An effect size of one, $\delta = 1$, indicates a two-fold change when the logarithm of the intensity is base 2. Different sequences on the microarray have different values of σ, as demonstrated in Figure 3.8. A conservative estimation for sample size is to use the value at the 90th percentile of the σ distribution. From Figure 3.8 for human tissues in two-color CpG island microarray experiments with common reference design, we find it to be around $\sigma \sim 0.5$ or $2^{0.5} = 1.4$ fold. $\alpha = 0.001$ gives on average ten false positives per ten thousand true negatives. (The choice of 0.001, instead of 0.05, has to do with multiple testing correction, which is discussed in section 5.5.) Equation (3.1) with $\alpha = 0.001$, $\beta = 0.1$, $\delta = 1$, $\sigma = 0.5$ then gives us the number of biological samples per class $n = 13$ needed to achieve a statistical power of 0.9 in detecting over twofold differentially methylated loci between the two classes with an average of ten false positives per ten thousand nondifferentially methylated loci on the microarray. Figure 3.9 to Figure 3.11 show the sample sizes for other settings of the parameters.

A reference design uses about twice the number of microarrays a balanced block design does because one of the channels of every microarray in the reference design is allocated to the same reference sample. Balanced block design, therefore, provides microarray savings. However, the variability σ in the log-intensity-ratios in a balanced block design is, in general, larger than that in a reference design since the ratio in a balanced block design is of

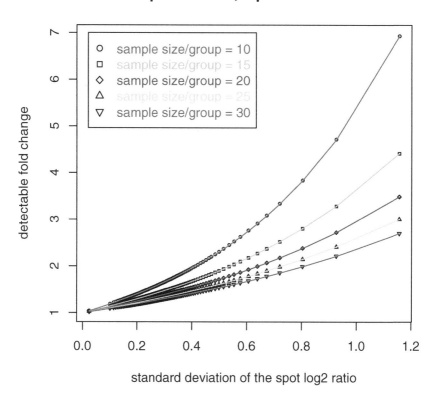

FIGURE 3.9: Minimum fold change versus variability in $\log_2$ ratios over different sample sizes at a given power and false positive rate.

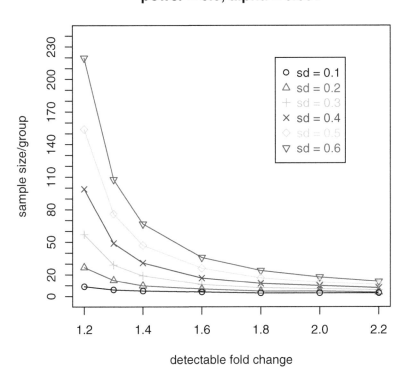

FIGURE 3.10: Sample size versus minimum fold change over various variability in $\log_2$ ratios at a given power and false positive rate.

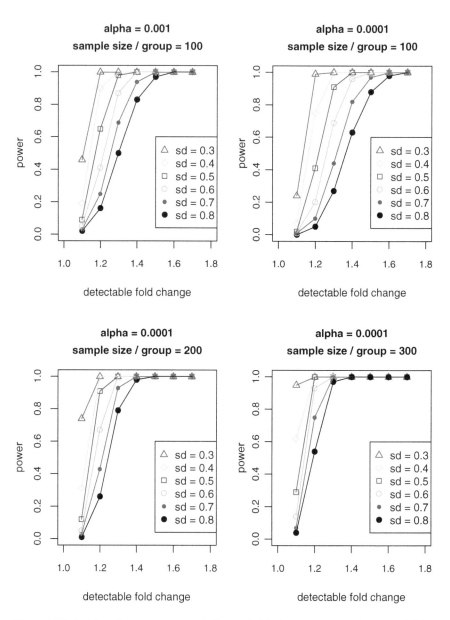

FIGURE 3.11:   Power versus minimum fold change over various variability in $\log_2$ ratios at two different sample sizes and false positive rates.

individual i to individual j, which tends to be more diverse than individual i to a common reference in a reference design. To achieve the same power as a reference design, a balanced block design, therefore, would need more biological samples, according to equation (3.1).

3.7.3 Pooling biological samples

A practice in microarray experiments is to mix independent biological samples into a pool prior to labeling and hybridization. The rationale behind pooling is that differences in DNA methylation (or gene expression) due to sample-to-sample variation are reduced by sample averaging [Kendziorski05]. The resulting reduction in the overall variance of the methylation at a locus σ^2 is

$$\sigma^2 = \frac{\sigma^2_{\text{bio}}}{R} + \frac{\sigma^2_{\text{tech}}}{M} , \tag{3.3}$$

where R is the number of samples in a pool and M the number of technical replicates per sample. Note that the within-class biological variability σ_{bio} of methylation varies wildly from locus to locus while the technical variability σ_{tech}, being more of a property of the instrumentation, is relatively constant over loci (cf. Figure 3.8). Pooling (i.e., $R > 1$) thus can be very effective for those loci whose σ_{bio} are much greater than σ_{tech}.

Since less microarrays will be consumed, pooling is advantageous when microarrays are expensive relative to samples. However, to be able to assess the within-population variation of methylation, we should not mix all samples together into a single pool. Instead, we arrange, say, fifteen samples from one class into three pools and fifteen samples from the other class into the other three pools, each pool consisting of five samples. Pooling in this way, in principle, is not to undermine identification of differentially methylated loci. If the number of microarrays per class is less than three, pooling is the only choice. The disadvantage of pooling is that specific information on individual samples is lost, preventing removal of outlying and thus likely poor quality samples.

In a large design involving many samples and microarrays, pooling extra samples onto a fixed number of microarrays decreases the variability across experiments. On the other hand, if we pool a fixed number of samples on less microarrays, we can risk finding less number of differentially methylated loci. For example, compare a measurement of ten samples using twenty microarrays with the other measurement of the same number of samples, but using only five microarrays. The former design has $R = 1$ and $M = 2$ while the latter $R = 2$ and $M = 1$ in equation (3.3). Suppose the biological variances σ^2_{bio} of loci g1, g2, g3 are, respectively, 4, 2, 1 (in an arbitrary unit, and same below), and their technical variances σ^2_{tech} are all equal to 2. The overall variances σ^2 of the three loci g1, g2, g3, according to equation (3.3), are, respectively, 5, 3, 2 in the former scenario ($R = 1$ and $M = 2$), and become, respectively, 4, 3, 2.5 in the latter scenario ($R = 2$ and $M = 1$). We see that, because of

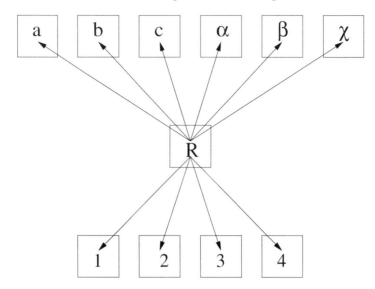

FIGURE 3.12:   Schematics of a reference design for three group comparison. Samples a, b, c are from one class. Samples α, β and χ are from the second class while samples 1, 2, 3 and 4 are from a third class. R is a common reference.

varying biological variability in DNA methylation, pooling while reducing the number of microarrays helps identify some differentially methylated loci (e.g., locus g1 in this example) at the expense of other loci (e.g., locus g3) albeit the sacrificed loci are not many according to the distributions in Figure 3.8.

We have discussed pooling in terms of class comparison. If we are also interested in class discovery by, for example, the method of clustering on methylation data, we do not want to pool samples. In fact, as the price of a microarray continues to drop, we do not need pooling except when scarcity of samples necessitates it.

3.8   Appendix

The reference, balanced block and loop designs in Figure 3.1 to Figure 3.3 are concerned with two-class comparisons. They were presented so because of ease of illustration not because of their limitation. Here, in Figure 3.12 to Figure 3.14, we extend the designs to the task of comparisons among three classes.

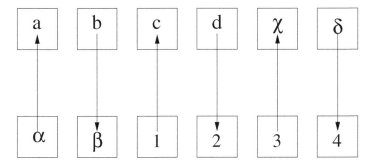

FIGURE 3.13:   A balanced block design for comparing samples from three groups. Group one samples include a, b, c, d; group two samples are α, β, χ, δ; and group three samples are 1, 2, 3 and 4.

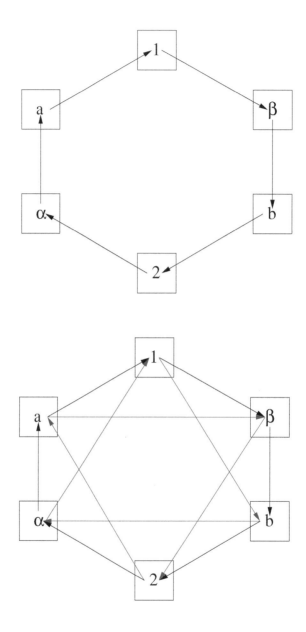

FIGURE 3.14: (Top) A loop design that involves samples from three groups: samples a, b are from a group; samples α, β from the other group and samples 1 and 2 from a third group. (Bottom) If we have more microarrays, a set of replicate arrays are the hybridizations shown in blue.

Chapter 4

Data Normalization

Normalization is a preprocessing procedure that is routinely applied to microarray data. The objective of normalization is to minimize, or even remove, the effects on data of artifacts that were inevitably introduced into the measurement due to imperfection of the technologies. Only after proper normalization can we contend with confidence that the identified changes in methylation or expression are of biological origins. Although nonbiological effects can be modeled together with biological ones in the analysis, we opt for a two-stage approach where methylation data are normalized first. Normalized data are then subject to exploration by various methods, which are the topics of subsequent chapters. Because of the two different types of microarrays, the ways they are normalized are somewhat different. Finally, in cases where the principles underlying the normalization are not well justified, we resort to the hybridization of known DNA fragments with the counterpart control probes on the microarray for normalization.

4.1 Measure of methylation

After hybridization in a microarray experiment, unbound nucleic acids are washed off the microarray slides. The quantity of the DNA sequences bound to a probe reflects the amount of either methylated or unmethylated DNA fragment (depending on which DNA fraction was enriched — methylated or unmethylated) in the sample. Recall that the sequences were labeled with fluorescent dyes prior to hybridization. A laser beam in a microarray scanner then shines on the probes to excite the dyes. Cyanine 3 (Cy3) labeled sequences will emit green light ($\sim$570 nm) while Cyanine 5 (Cy5) red light ($\sim$670 nm) upon excitation. The emitted photons are then measured by a light sensor, such as photomultiplier tube (PMT) or CCD camera in the scanner. The scanning produces an image file, usually in uncompressed TIFF format, for each color. An image analysis software is then used to locate the bright spots in the TIFF image and quantify the brightness, as well as the quality, of the spots. Results of the image analysis are output to a file of plain text format, such as `my_array_data.gpr` and `my_array_data.spot` for two-color DNA microarrays or `my_array_data.cel` for single-color oligonucleotide chips. The remaining sections are dedicated to the analysis of the image-analyzed data.

In the file, results of image analysis are usually rendered in a spreadsheet. A row corresponds to a probe sequence. Different columns display different attributes of the sequence, such as its ID, the physical location of the spot on the microarray; the mean, median and standard deviation of the intensities from the foreground pixels (due to the target sequences and are called signal or foreground intensities); the mean, median and standard deviation of the background intensities (due, for example, to stray fluorescence from other chemicals on the slide); and other quality control measures of the spot intensities, such as signal to noise ratio, circularity of the signal intensities, and so on. The meaning of each column is manifested in the header of the table. Magnitudes of the intensities are between 0 and 65,535. That is, they are 16-bit integers resulting from the 16-bit grayscale TIFF image acquired by the scanner.

Microarray measurements are noisy; the levels of random noise we have experienced are in the 10 to 40 percent range. The number of fragments interrogated in a microarray can be hundreds to tens of thousands. The number of microarrays hybridized in a single experiment can be a dozen to over a hundred. In view of the large quantity of noisy data from microarray experiments, we need statistics to help better analyze the data and interpret the result.

Many powerful statistics tools have been developed for Gaussian (i.e., normal) distributions. The ubiquity of normal distributions is ascribed to the central limit theorem. We, therefore, first of all would like to have a look at the distributions of the green and red intensities of the spots in a microarray. Figure 4.1 shows examples of such raw intensity distributions from two dual-color microarrays by two different scanners. We see that the distributions are skewed with long tails to the right. The skewed distributions suggest a model for the methylation that consists of an exponential component plus a Gaussian noise,

$$y = \eta e^{\mu} + b + \epsilon \,, \tag{4.1}$$

where y and b are, respectively, the green (or red) signal and background intensities recorded in the data file, μ is the methylation level of the locus in the Cy3-labeled (or Cy5-labeled) sample, η is a multiplicative constant, and ϵ is a Gaussian noise of the background with mean zero and standard deviation σ. With the model of equation (4.1), we take logarithm of the background-subtracted intensity to get the methylation level μ,

$$\log(y - b - \epsilon) = \mu + \log \eta \,. \tag{4.2}$$

In Figure 4.2, we show again the distributions of the data of Figure 4.1, but the intensities in the distributions are now background corrected and logarithmically transformed. We see that they become more symmetric and closer to Gaussian distributions. Most of the tools of classical statistics then can be

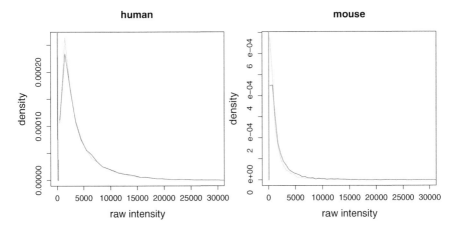

FIGURE 4.1: Distributions of raw signal and background intensities for a 13,056-spot human CpG island microarray on the left and 7680-spot mouse CpG island microarray on the right. The vertical green and red spikes near zero are the backgrounds.

readily applied to the logged microarray data. The other advantage of logarithmic transformation of the intensities is that effects that are multiplicative become additive (cf. η in equation (4.1) and equation (4.2)).

4.2 The need for normalization

A microarray experiment involves many steps, including DNA or RNA isolation, restriction enzyme digestion and PCR amplification for methylation, labeling, hybridization, washing, and scanning. Any of these is prone to random noise and systematic biases. Even when the same experiment with samples from the same biological source, reagents of the same batch, equipments in the same laboratory, is repeated by the same experimenter following the same protocol at the same ambient conditions, the measured intensities are bound to be different. This technical variability is an example of random fluctuations. On the other hand, it has been known that Cy3 fluor is incorporated to nucleotides with a higher efficiency than Cy5. The measured green light intensities will be higher than red light even though the same amounts of Cy3- and Cy5-labeled target sequences are admitted to the mixture for hybridization. This asymmetric dye effect is an example of systematic bias.

As an illustration of the obscuring effects of random and systematic errors, we show in Figure 4.3 examples of the so-called self hybridization where two

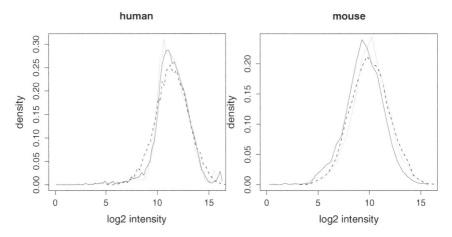

FIGURE 4.2:   Distributions of the background-subtracted intensities. Data are from Figure 4.1. The dashed-line distributions on each side come from 13,056 and 7680 draws from the normal distributions whose means and standard deviations are calculated from the corresponding intensity distributions.

aliquots from a mixture are separately labeled with Cy3 and Cy5 fluors for co-hybridization. Since the Cy3 and Cy5 channels are derived from an identical cellular source, we would expect a straight line along the diagonal of a green versus red intensity scatter plot. Deviations from the diagonal clearly manifest a need for normalization.

4.3   Strategy for normalization

To accommodate random and systematic noise, we can extend equation (4.2) by adding, to the right-hand side of it, a noise term and as many bias terms as desirable. The resulting model is linear additive. Many regression methods in statistics then can be utilized to estimate the parameter of interest, i.e., μ, along with other nuisance effects. This one-step approach is straightforward. However, in many cases estimating methylation (or expression) levels is not the end of the story. We would like to pick up differentially methylated spots. We may want to categorize sequences according to their methylation profiles by the use of clustering methods. We might further want to investigate the interwoven dependences (interactions) among sequences (genes). The various analyses are most likely implemented in different application software on different platforms and explored by different people of diverse research inter-

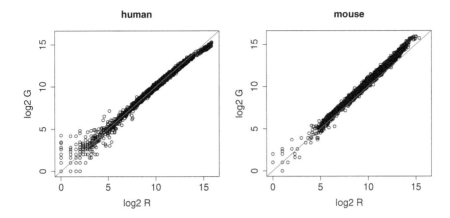

FIGURE 4.3: Log$_2$ red versus log$_2$ green intensities from self hybridizations. All spot intensities are background-corrected. (Left) Human blood DNA on a human CpG island microarray. (Right) Mouse brain DNA on a mouse CpG island microarray.

ests. A general strategy, therefore, is to estimate the methylation (expression) without tying oneself to the detailed assumption on the error model. We call this a two-step process where nonbiological artifacts are firstly eliminated to a great extent from the data. The normalized data then are exported for further analysis [Kerr01b, Wolfinger01].

As we will see, normalization itself is further modularized. Modularization lends flexibility. We are free to justify the various combinations of normalization options before plunging into the next phase of analysis.

4.4 Two-color CpG island microarray normalization

Two-color DNA microarrays are among one of the popular microarrays. Their flexibility accounts, in part, for the prevalence. In fact, many homemade microarrays belong to this technology category. Also an important feature of DNA microarrays is that they perform direct comparisons between case and control samples by forming methylation (expression) ratios. Because of the ratio, many within-array artifacts, such as the effects due to variations in spot size, spot shape, hybridization conditions, etc., are canceled. A major concern about DNA microarrays is nonspecific hybridization. The number of bases of the single-stranded DNA that are immobilized on the spot in a DNA microarray is in the hundreds (∼500 bp). The longer the single-stranded DNA molecule, the more susceptible it is to cross hybridization.

4.4.1   Global dependence of log methylation ratios

The number of probes in a microarray is huge, reaching tens of thousands. Selection of CpG islands or genes as probes is random for pioneering studies. Even for a complete coverage of the genome on the microarray, the numbers of genes that undergo over- or under-expression under an experimental setting are not thought to be as huge. For example, in a microarray experiment to search for genes that predispose to a disease, we would expect to find only few such aberrant genes, plus the secondary genes that are regulated by the aberrant genes. The rest of the genes behave normally. The expression levels of the normal genes in the diseased samples should be the same as those of the normal genes in the healthy samples. Similar arguments apply to DNA methylation.

Diseased samples are conventionally labeled with red fluor and healthy samples with green. Using our measure of methylation (or expression) in section 4.1, we calculate the difference in the log red intensity, R, and log green intensity, G, of a spot, $\log R - \log G = \log(R/G)$. It is understood that all intensities are background subtracted before log transformation. Note that in microarray community, base 2 logarithm is more often used than the natural logarithm because $\log_2(R/G) = 1$ spells that the red intensity is $2^1 = 2$-fold over that of the green intensity.

Since most loci show no differential methylation, the majority of the log intensity ratios (or simply log ratios) are zero and the distribution of these log ratios should peak around zero. Sequences whose log ratios are far away from the peak are called outliers. Figure 4.4 shows examples of the distributions of log ratios from self hybridizations. There are indeed peaks in the histogram, but the centers of the peaks are off zero. When we plot more such histograms from other two-color microarrays we observe that the peaks usually sit on the left-hand side of zero. This again demonstrates a low incorporation efficiency of the red (Cy5) dye in comparison to the green (Cy3) dye. Furthermore, different PMT (photomultiplier tube) high voltage settings during scanning also can shift the peak.

A first step toward correcting for the dye bias is thus to bring the center of the distribution to zero. This is easily done by subtracting the log ratio of each sequence i by the median of the log ratios,

$$\log_2(R_i/G_i) = \log_2(R_i/G_i) - c ,\qquad (4.3)$$

where c is the median of the log intensity ratios. Recall that median is less influenced by extreme values than mean, which explains why it is used in the correction. The = sign in equation (4.3) means assignment; the center-adjusted log intensity ratio on the right is assigned to and replacing the origi-

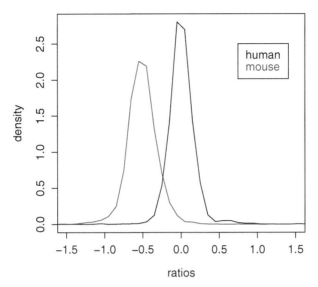

FIGURE 4.4: Distributions of log intensity ratios from self hybridizations. (Data are from Figure 4.3.)

nal log intensity ratio on the left. This is the syntax used in most programming languages and environments including R and Bioconductor (cf. chapter 13).

4.4.2 Dependence of log ratios on intensity

The scatter plot of red versus green intensities as in Figure 4.3 reveals artifacts in a microarray measurement. In a two-color DNA microarray, the quantities of primary interest are methylation or expression ratios (i.e., fold changes), which are numerically obtained by $\log_2(R/G) = \log_2 R - \log_2 G$. In the spirit of Figure 4.3, it would be informative if we can have a scatter plot that uses log intensity ratio as one of the axes. A natural choice for the other axis would be $(\log_2 R + \log_2 G)$, which is independent of or, in mathematical jargon, orthogonal to $(\log_2 R - \log_2 G)$. We decide to use $(\log_2 R + \log_2 G)/2$, which has an additional property of being the average of the log red and log green intensities. The resulting scatter plot is call M-A plot [Dudoit02] where M stands for minus and A for average,

$$
\begin{aligned}
M &= \log_2 R - \log_2 G \\
A &= \tfrac{1}{2}\left(\log_2 R + \log_2 G\right)
\end{aligned}
\tag{4.4}
$$

Figure 4.5 shows the M-A plots of Figure 4.3 after the global correction by median equation (4.3).

Recall that we would expect a vast throng of zeros with only a meager presence of outliers. The zeros should be independent of the fluorescent intensities. That is when the $\log R$ value of a spot is large, its $\log G$ value is

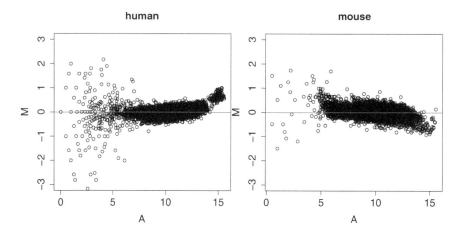

FIGURE 4.5:   Log intensity ratio M as a function of average log intensity A. Fluorescence intensity data are from Figure 4.3 but transformed according to equation (4.4).

also large so that their difference remains close to zero. An M-A scatter plot should show a cloud of points along the horizontal line at M = 0 across A's. The banana shape of points in the M-A plot of Figure 4.5 rebuts the expectation. This discloses the artifact of dependence of ratios on intensities. To correct for such intensity dependence, we can extend equation (4.3) such that the constant c now becomes a function of the average intensity,

$$\log_2(R_i/G_i) \leftarrow \log_2(R_i/G_i) - c(A_i) . \tag{4.5}$$

It would be helpful if we could find the $c(A)$ with an analytical form that is universal to microarrays. Unfortunately, when we plot more and more M-A plots from different microarrays, we find persistence of the artifact, but the shape can be quite different from one microarray to another. Figure 4.5 already shows two hybridizations and the shapes are indeed different. A handy algorithm called loess (or lowess for robust locally weighted polynomial regression) in statistics comes to the rescue. Loess first breaks the distribution of data points into segments and then fits a linear line or quadratic curve to each segment of the data. The algorithm in essence works by representing a function of complex form by pieces of simple linear or quadratic functions. Results of loess fitting do not depend on the small percentage of extreme values in the dataset. Furthermore, it is robust as long as the number of data points in each segment is large enough.

4.4.3 Dependence of log ratios on print-tips

A microarray is found to have a dependence of M on A of its own. We know that spots are arranged in grids on the surface of a microarray slide. The spots are printed by a set of print-tips, each of which prints a grid of spots. Because of the peculiarity of each print-tip, it is not surprising to find that the spots by a print-tip have a unique dispersional distribution of M versus A pertinent to the tip. We show in Figure 4.6 and Figure 4.7 the M-A plot and M-value boxplot of Figure 4.5 stratified by different print-tips.

To correct for the print-tip effect, we extend equation (4.5) so that the correcting function $c(A)$ depends not only on intensity, but also on print-tip [Yang01, Yang02, Smyth03],

$$\log_2(R_i/G_i) = \log_2(R_i/G_i) - c(A_i, \text{Tip}), \tag{4.6}$$

where Tip is the print-tip group to which probe i belongs. The normalization of equation (4.6) is no harder than repeating the loess regression equation (4.5) on the spots belonging to a print-tip as many times as there are different print-tips that printed the microarray. Information on the print-tip layout is usually stored in the same data file containing the raw intensities. We show in Figure 4.8 and Figure 4.9 the M-values boxplots and M-A plots of the data normalized by the print-tip loess method equation (4.6).

In designing a microarray, the spot location for the probe sequence on the array is chosen at random. That is, it is unusual, if not unwise, to print those protein synthesis-related genes in one grid or block of the microarray, metabolism related genes in another block, and so on. Similarly, chromosome six CpG islands are printed all over the microarray instead of concentrating in one or few grids. As a consequence, the spreads (i.e., standard deviations or median absolute deviations) of the distributions of intensity ratios by different print-tips should be more or less the same. Boxplots of the normalized log intensity ratios by different print-tips are shown in Figure 4.8. The horizontal bar inside the box is the median (i.e., 2nd quartile) of the distribution. The lower and upper edge of the box enclose, respectively, the 1st and 3rd quartile of the data. The length of the whisker is by convention 1.5 times the height of the box. The data points beyond are plotted because they are likely outliers. The taller the box, the wider the distribution, indicating that many of the sequences printed by the print-tip are hyper- or hypomethylated (over- or under-expressed).

Since spot locations are chosen randomly, it is unlikely to find nonuniform box heights from one print-tip to another. If it does happen, the cause of such systematic bias can be unequal lengths or openings of the print-tips. One remedy is to multiply a constant factor to each ratio of the same print-tip. That is, shrink (say, $\times 4/5$) the over-estimated ratios from one print-tip and inflate ($\times 5/4$) the under-estimated ones from the other so that the spreads of all the ratios are roughly the same, independent of the print-tips [Yang02]. Note that this scaling factor method would increase the overall

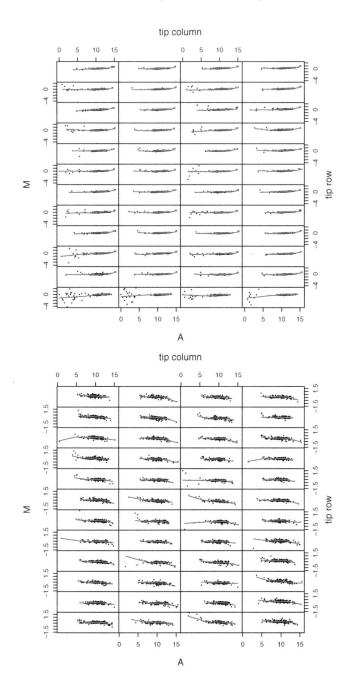

FIGURE 4.6: The M-A plot of Figure 4.5, but stratified according to the print-tips on the microarrays (top: human, bottom: mouse). The red curve shows the loess fit to data points in a grid on the microarray.

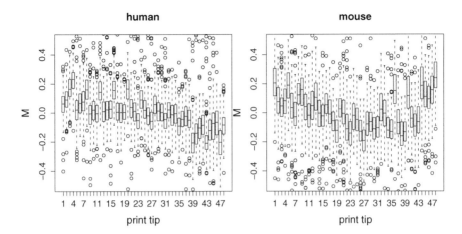

FIGURE 4.7:   Boxplots of the print-tip group M-values in Figure 4.6.

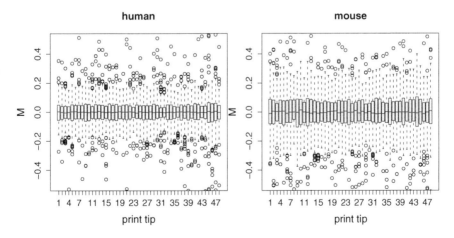

FIGURE 4.8:   Boxplots of the print-tip group M-values after the print-tip loess normalization of equation (4.6).

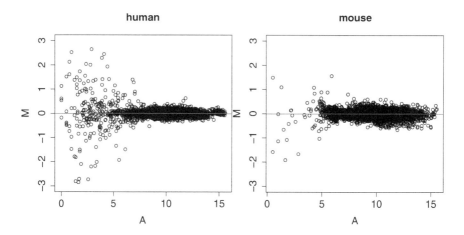

FIGURE 4.9:   M-A plots of the print-tip loess normalized data (cf. Figure 4.5).

standard deviation (cf. Appendix (section 4.7)). It should not be used without justification. In fact, as the microarray technology advances, there may be no need for such correction.

4.4.4   Normalized Cy3- and Cy5-intensities

In some applications, we are interested in normalized Cy3- and/or Cy5-intensities. That may happen when, for example, the Cy3 channel is associated with samples from twins and the Cy5 channel with those from the co-twins. It might then be more convenient to work with $\log_2 R$ and $\log_2 G$ than the fold change. We can solve $\log_2 R$ and $\log_2 G$ in terms of M and A from equation (4.4),

$$\log_2 R = A + \tfrac{M}{2}$$

$$\log_2 G = A - \tfrac{M}{2} \, .$$

(4.7)

The normalized $\log_2 R$ and $\log_2 G$ are obtained by substituting the print-tip loess adjusted M-value equation (4.6) into the above formula. Figure 4.10 shows distributions of the normalized log intensities of Figure 4.2.

Note that we cannot treat such normalized single color ($\log_2$) intensities from a two-color DNA microarray as two independent measurements of the samples. They are not fully independent because the two measurements share the same make of the spot, hybridization condition, ..., etc. The intensities from the two channels are highly correlated. $\log_2 R$ and $\log_2 G$, therefore, do not give us much more information than the ratio $\log_2 R/G$ alone. In other words, the nature of competitive hybridization between Cy3- and Cy5-

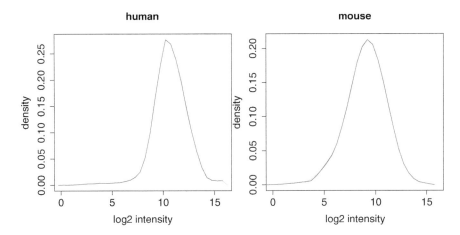

FIGURE 4.10: Normalized $\log_2$ intensities of Figure 4.2.

labeled sequences to the same probe makes two-color DNA microarrays good at measuring methylation (or expression) difference.

4.4.5 Between-array normalization

The normalization procedures we have so far dealt with are within-array. After within-array normalization, artifacts are removed as much as possible. We could hopefully proceed to look for differentially methylated spots. However, in light of the substantial noise in microarray technology, a single measurement speaks little. To be able to build credibility of the result, we need to replicate the measurement. Statistical tools can then be used to assess the confidence in the result before we report a finding.

Replication can be of different levels. When the biological samples in the replicates are derived from the same cells, we call these technical replicates. In this case, since the samples in the microarrays have identical genotype and phenotype, after within-array normalization, we would expect the spreads of the ratios to be approximately the same across microarrays. Nonuniformity in the spreads suggests evidence for variation in, e.g., hybridization. If this is the case, we can apply the scaling factor technique in the previous section to perform between-array normalization. Separate constants are multiplied to individual microarrays, bringing the spreads of ratios distributions to a common value. The between-array normalization ensures that class (or group) means are not dominated by one or few outlier microarrays.

In most applications, we hybridize DNA samples from independent biological sources to microarrays. For example, we want to study the molecular etiology of a disease. We then collect samples from many affected individuals for microarray assay. This is an example of biological replicates. Because of idiosyncrasies, we would not expect the expression profiles to be the same across individuals. The assumption of a uniform spread across microarrays no longer holds. We need other criteria, such as control spots, for between-array normalization in such cases.

4.5 Oligonucleotide arrays normalization

Oligonucleotide chips, such as Affymetrix® promoter arrays, are among one of the most popular microarray platforms. Oligonucleotide tiling arrays are discussed in chapter 6. They generally provide a larger number of probes per unit area than two-color DNA microarrays. The experimental design for oligonucleotide chips is relatively simple because hybridization involves only one sample. A second sample will have to be hybridized to a second chip. (Pooling of samples was discussed in section 3.7.3.) Comparisons made between the samples are then indirect. The major disadvantage of oligonucleotide chips lies in their relatively high cost.

4.5.1 Background correction: PM – MM?

In oligonucleotide technology, the nucleotides in a probe are typically 25-mer. They are synthesized on the surface of the chip by photolithographic techniques developed by Affymetrix, Inc. A similar platform was also developed by Agilent Technologies, Inc.® The typical number of bases in a gene or CpG island is definitely greater than twenty-five. Therefore, in the design, hundreds and thousands of probes along the DNA segment of interest are chosen to represent the gene or CpG islands. The probes are called perfect match (PM) because their DNA sequences are from verified public databases. To be able to detect nonspecific binding, the thirteenth base of a perfect match is deliberately mistaken in making a so-called mismatch (MM) probe. A PM probe and its corresponding MM probe form a probe pair. Subtraction of MM intensity from PM intensity, in principle, gets rid of signals from cross hybridization. The set of eleven to twenty probe pairs is called a probeset, which collectively measures the methylation status or expression level of a target sequence.

The idea of using mismatches to ward off nonspecific hybridization is neat. However, it was found that MM intensities increased with PM intensities, suggesting that mismatch probes measure not only nonspecific hybridization, but also specific hybridization. It is also known mathematically that subtraction elevates uncertainty. For example, differencing increases the variance by a factor of 2. It, therefore, is contended that background correction by subtracting MM from PM not be performed if the gain does not outweigh the

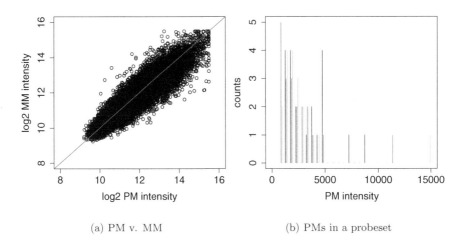

(a) PM v. MM   (b) PMs in a probeset

FIGURE 4.11: (a) Scatter plot of PM versus MM $\log_2$ intensities on an oligonucleotide chip that contains 835,396 probes and 16,726 AffyIDs. Points above the red line indicate MM > PM. (b) The PM intensities from a probeset. Each color is from an oligonucleotide chip. Note that histograms from different chips are slightly displaced for displaying purpose.

loss. Moreover, it is not rare to find a probe pair with MM intensity > PM intensity (cf. Figure 4.11(a)). Subtraction would lead to a negative methylation or expression that inflicts conceptual difficulty. Therefore, in case of doubt, the reader can simply focus on the PMs and leave alone the MMs without losing much. Figure 4.11(b) shows an example of the PM intensities from the same probeset on three oligonucleotide chips.

4.5.2 Quantile normalization

The number and coverage of sequences in a high density oligonucleotide chip reaches the genomic scale. In a mutant versus wild type experiment, one or two genes are knocked out in the mutant mice. The number of affected genes due to the knockout is assumed to be only a tiny fraction of the genome. The bulk of the genome expression is intact. Provided we can make the assumption, the distributions of the PM intensities across different chips should be very much alike. Methylation distributions are also conditioned on the proviso.

Statisticians use quantile-quantile plot (qqplot) to visualize if two distributions are the same. If they are, their quantiles should delineate a straight line along the diagonal of the plot. The quantile normalization for oligonucleotide chips is an extension of the two-dimensional qqplot to N dimensions where

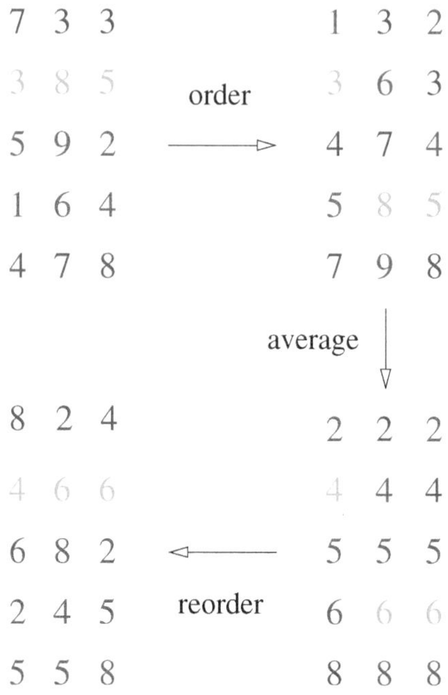

7 3 3

3 8 5   order

5 9 2   ⟶

1 6 4

4 7 8

         1 3 2

         3 6 3

         4 7 4

         5 8 5

         7 9 8

average ↓

8 2 4

4 6 6

6 8 2 ⟵

2 4 5 reorder

5 5 8

         2 2 2

         4 4 4

         5 5 5

         6 6 6

         8 8 8

FIGURE 4.12:   The three steps in quantile normalization. Log PM intensities on a chip are stored in a column. (The pen color helps place the averaged
numbers back to the original positions.)

N is the number of PMs on a chip [Bolstad03]. The algorithm is simple and
fast. Usually we start by storing the PM intensities (or PM and MM if the
mismatch is decided to be accounted for) in a matrix with different rows for
the different PM readings and columns for chips. The quantile normalization then consists of three steps: (1) sort the numbers in each column in an
ascending order, (2) replace each number in a row with its row mean, and
(3) rearrange the numbers in the column into its original order. Figure 4.12
illustrates the algorithm with a toy system containing three chips each having
five PM probes. Figure 4.13 shows the distributions of $\log_2$ PM intensities
from three oligonucleotide chips with and without quantile normalization.

Note that the assumption behind the quantile normalization is very stringent. In the case where the phenotypes and/or genotypes (i.e., SNPs) among
the samples are known to be very different, we may be hesitant to assume a
common shape in the distributions. The numbers of sequences in high density oligonucleotide chips are nevertheless huge. The assumption that most of
the methylation (expression) are invariant across comparisons may still hold.
Because it is also the assumption on which the two-color DNA microarray

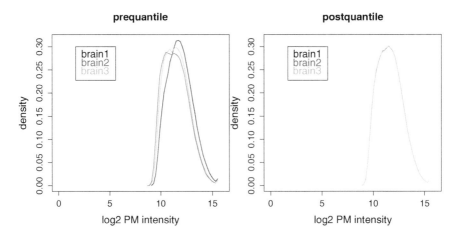

FIGURE 4.13: Distributions of the log₂ PM methylation intensities from three postmortem brains on three oligonucleotide chips before quantile normalization on the left and after on the right.

normalization is based, we can borrow the methods for two-color DNA microarrays from section 4.4.1 and section 4.4.2. In doing so, we first form pairs of oligonucleotide chips so that one chip in the pair measures red channel while the other green. We then proceed as though they were two-color DNA microarrays. Namely we apply equation (4.5) on the probe intensities from each pair of the chips. After the first round exhausts all possible pairings, we go for the second round. We continue the cycle until there is little adjustment in the intensities between successive iterations. The procedure can become computationally intensive especially when the number of chips in the experiment is large.

4.5.3 Probeset summarization

Methylation or expression levels are our quantities of concern. The PM values in a probeset can differ from one another by two orders of magnitude (i.e., 100 times, cf. Figure 4.11(b)). We need a mechanism that summarizes the eleven to twenty PM values in a probeset into a single measure of methylation status (expression level) for the sequence.

Two-color microarrays and oligonucleotide chips take advantage of the preferential binding between adenine and thymine and between guanine and cytosine in the hybridization. Both capture the light emanating from the fluorescent dye. They share many commonalities in the technology. It is not surprising the methylation (expression) levels by oligonucleotide probesets can be modeled in a similar way. Following equation (4.2), for a probeset, the methylation (expression) level μ appears linear in the logarithm of the

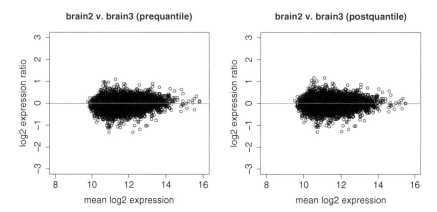

FIGURE 4.14: Log methylation ratios of brain 2 to brain 3 as a function of the average log methylation before (left) and after (right) quantile normalization. (cf. Figure 4.13.)

individual PM values,

$$\log_2 \mathrm{PM}_j = \mu + \alpha_j + \epsilon_j \,, \tag{4.8}$$

where α accounts for the effect of probe such as affinity, ϵ is a Gaussian noise with mean zero, and j indexes the probe inside the probeset [Irizarry03]. With a constraint like $\sum_j \alpha_j = 0$, μ can be readily estimated by a robust regression method, such as median polish, in statistics. In Figure 4.14, with the probeset summarization equation (4.8), we show the pre- and postquantile normalized M-A plots of one possible comparison of methylation between the three oligonucleotide chips in Figure 4.13.

4.6 Normalization using control sequences

The normalization has so far been based on very general assumptions on the distribution of the differential methylation and transcription. Namely, an overwhelming majority of the DNA fragments in a two-color DNA microarray comparing, say, the diseased with healthy groups of samples exhibit no change in methylation and expression. Or, the locations and shapes of the PM distributions from the many oligonucleotide chips in an experiment are virtually identical. Some research laboratories with specific target sequences in mind produce their own homemade microarrays with only a limited number of preselected sequence fragments on them. Examples can include SARS (severe acute respiratory syndrome) microarrays that interrogate only the genes involved in the respiration and related pathways. The CpG islands around a few selected genes, such as tumor suppressor genes, can also be spotted on a boutique microarray for cancer diagnosis. (We will not address custom oligonucleotide chips because the production cost can be too high for most to afford.) The assumption breaks down in such cases.

In order to keep using the normalization algorithms that have been developed, one way is to utilize the sequences whose methylation (expression) levels are known to be invariable under the varying biological contexts studied in the experiments. Genes that code for proteins involved in such basic cellular maintenance functions as cytoskeletal component, glycolytic pathway, protein folding, ribosomal synthesis are traditionally selected as control or so-called housekeeping genes. The CpG islands in the promoter regions of the housekeeping genes thus are unmethylated irrespective of samples, serving as negative controls. In addition, control sequences are also chosen from DNA exogenous to the genome under study. They also can be artificial sequences that share no significant sequence homology with the eukaryotic and prokaryotic genomes sequenced so far. In these cases, polymerase chain reaction (PCR) products of the control sequences are provided for microarray printing. The associated mRNA spikes are also commercially available for reverse transcriptase labeling. Recall that the intensity ratio depends on the intensity in a bizarre way (cf. section 4.4.2). Various amounts of mRNA spike-ins to the fluorescent labeling, therefore, should be prepared to cover the full range of hybridization intensities. Note that spikes for the Cy3- and Cy5-labeling reactions can also be prepared at predefined abundance ratios, serving as standardized positive controls for the purpose of calibration and between-array normalization.

Given the negative control spots with methylation or transcription properties that meet our normalization assumptions, we run the loess normalization equation (4.6) through them. That is over only those indexes is that represent the negative control spots. Once we obtain the the correction function $c(A, \text{Tip})$ this way, we repeat the process equation (4.6), but now over the rest of the spots on the microarray [Yang02, Smyth03]. This procedure equivalently sets zero weights to spots other than the negative control spots in determining the correction. Figure 4.15 and Figure 4.16 show two examples of the normalization with control spots on a custom dual-color CpG island microarray containing a total of only four hundred spots. In these examples, log intensities of all the spots were used in the print-tip loess, but we took advantage of the control spots by doubling their weights in the loess normalization.

Because of the local dependence of fold-change on intensity, if the spike-in DNA fragments or mRNA are absent at some concentration in the mixture, the correction $c(A, \text{Tip})$ will be missing its value at that intensity. As a consequence, we have to be cautious of any spurious results of normalization on customary microarrays in which the amounts of spike-in controls failed to cover the whole dynamic range of the hybridization intensities.

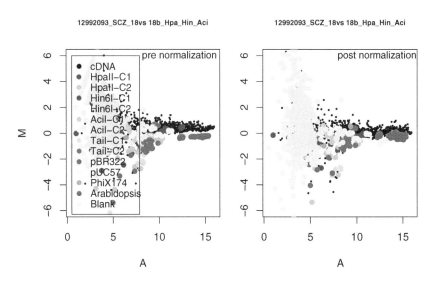

FIGURE 4.15: Normalization using the "control spots" (in color) on a CpG island microarray that consists of only four hundred spots.

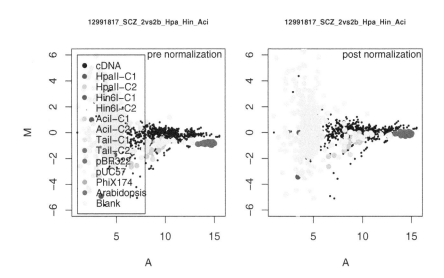

FIGURE 4.16: Normalization using the "control spots" (in color) on a CpG island microarray that consists of only four hundred spots.

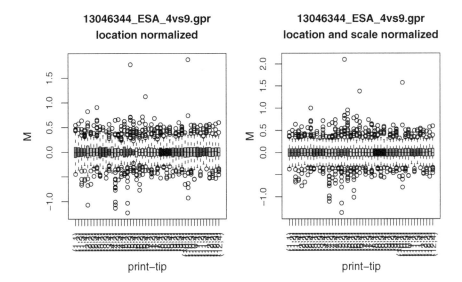

FIGURE 4.17:   Distribution of the log ratios across print-tips in a microarray without (left) and with (right) scale normalization.

4.7   Appendix

Figure 4.8 shows the normalization by separate loess regressions to the spots belonging to different print-tips. We mentioned in section 4.4.3 a scale normalization that brings the boxes to a common height. Figure 4.17 shows a scale normalization within a microarray. Remember that such an operation may increase the overall variation of the ratios. If the distribution of box heights are more or less uniform, such as in Figure 4.17, a scale normalization may not be necessary.

Figure 4.18 is a step-by-step illustration of the within-array and between-array normalization (for two-color CpG island microarrays) in action. Notice that array # 3 is penalized by the scaling in the between-array normalization. The use of scaling thus has to be justified.

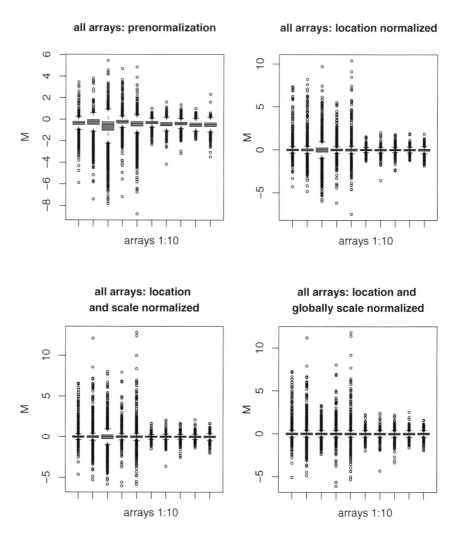

FIGURE 4.18:   Upper left: raw data. Upper right: print-tip loess normalization. Lower left: plus scale normalization within arrays. Lower right: plus scale normalization across arrays.

Chapter 5

Significant Differential Methylation

In mammalian genomes, many of the high CG-dinucleotide content regions occur in or near the promoters of genes. The so-called CpG islands are mostly unmethylated. CpG island methylation is correlated with gene expression silencing. Aberrant methylation thus was found to be associated with many developmental disorders and cancers. Identification of differential methylation between cases and controls has implications in molecular etiology, diagnosis and therapy.

An organism's genome consists of hundreds to many thousands of genes. To conserve resources, not all genes are necessarily expressed at all time. Instead, a minimum number of core genes are expressed, producing proteins for routine operation and maintenance of the cell. Other genes are expressed during various stages of cellular development and in response to dynamic environmental stimuli.

There, in principle, can be as many CpG island probes on a microarray for DNA methylation as there are genes in the genome. Microarrays are a high-throughput technology. One of its major applications lies in identifying the DNA regions or genes that exhibit changes in methylation or expression across different cellular or external conditions. Because of noise in the measurement and variability in the samples, we need statistics to help make inference about population from our experiment that involves only a limited number of samples. In later sections of this chapter, we also cover topics that go beyond differential methylation or expression. In discussing analysis of DNA methylation or gene expression, we assume the microarray data have already been properly normalized by the methods in the last chapter.

5.1 Fold change

If the average fluorescent intensity of the amplicons from affected samples is 1600 and that of the control samples is 800, we speak of a methylation (or unmethylation depending on the protocols used) fold change of 2 ($= 1600/800$). Methylation levels, which are presumably proportional to the methylated CpG sites in the probe, can go either up or down relative to the control. For hy-

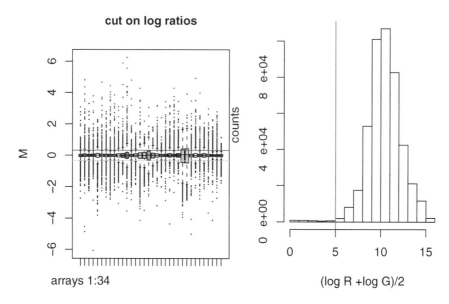

FIGURE 5.1:   Thresholds to the log ratios (left) and to the average log intensities (right).

permethylation, the range of fold changes is between 1 and infinity; while for hypomethylation, it is bounded between 1 and 0. The distribution of fold changes is not symmetric. Bioinformaticians therefore take the (base 2) logarithm of the hybridization intensity as the measure of DNA methylation or gene expression level. Distributions of log intensity ratios (i.e., log fold changes) of the thousands of probe sequences on a microarray then look like a Gaussian distribution, which is symmetric around zero. A quick look at the data for differential methylation is then to place a cut-off at, say, 2 on the positive side of the log ratio distribution for DNA fragments which underwent over four $(= 2^2)$ fold up methylation and -2 on the negative side for fragments which underwent over four $(= 2^{-2} = 1/2^2)$ fold down methylation in the microarray experiment. Figure 5.1 shows an example of thresholds to log ratios.

Normalized microarray data is known to show a characteristic funnel distribution in its plot of log ratios versus average log intensities (cf. the M-A plot of Figure 4.9 in section 4.4.3)): Low log intensities tend to vary wildly in their log ratios. Poisson statistics tell that the lower the amplicon (or mRNA) count, the larger the error in the proportion. A similar account says that measurements of weak signals are less reliable. Some researchers, therefore, remove spots of weak intensities prior to analysis (cf. Figure 5.1).

Because a microarray experiment almost always involves biologically and/or technically replicated measurements, we can calculate the standard deviation

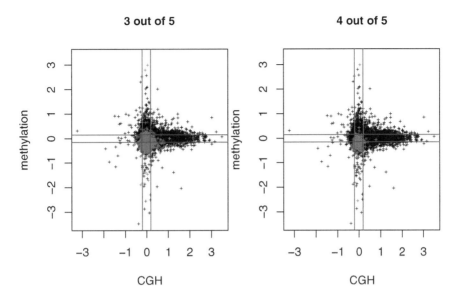

FIGURE 5.2: Spots whose log fold changes are over a threshold in at least three (left) and four (right) among the five measurements. CGH for comparative genomic hybridization measures gene copy number variations.

of the log ratios for every probe. Standard deviation, average deviation of the log ratios from the mean log ratio, serves as a measure of consistency of the measured log ratios from the replicated measurements. We can, in addition to a threshold to fold changes, put an upper bound on the standard deviations in identifying differentially methylated sequence fragments. Figure 5.2 shows a similar method that requires that the methylation fold change at a locus pass a certain threshold in at least three or four out of five measurements.

A threshold to fold changes, however, can be very subjective. One investigator may set a threshold at three-fold, while the other may decide that four-fold is the right value. If the threshold value is set too high, lots of spots of potential interest may be missed. On the other hand, if it is set too low, many uninteresting spots will be included for further investigation. The same criticism applies to bounds on standard deviations. We need a measure of *significance* for differential methylation. Note that the discussion applies to gene expression microarray data.

5.2   Linear model for log-ratios or log-intensities

A researcher has in mind a model describing a phenomenon of interest. She then sets up an experiment to test the adequacy of the model. If the results of repeated measurements fall within model predictions, say, 95 percent or more of the time, the agreement can be considered significant, lending her confidence in the model. In many disciplines, it is the so-called null hypotheses, or null (or reduced) models, that are conveniently, and thus traditionally, tested by experimenters. If results of the repeated measurements meet predictions of the null model 5 percent or less of the time, evidence for the null hypothesis is weak, prompting us to reject the null model.

Logarithmic transformation of fluorescent intensities symmetrizes the resulting distribution. The transformation also stabilizes the variability in the sense that magnitude of variation in the intensities is compressed by taking the log. Furthermore, thanks to the equality $\log(xy) = \log x + \log y$, multiplicative effects, such as fold change, become additive in the logarithmic space. Linear additive models thus have been developed for the log ratios from two-color DNA microarrays and the log intensities from single-color oligonucleotide chips. The models, being linear, are simple, yet they have been shown to represent a good approximation to the otherwise complex and even intractable phenomena, such as gene transcription [Kerr01b, Smyth04]. In what follows, we take a common reference design of two-color DNA microarray experiment as an example of linear model analysis for differential methylation and expression. The statistical analysis in this example applies equally well to data with single-color oligonucleotide chips.

5.2.1   Microarrays reference design or oligonucleotide chips

Suppose eight independent biological samples, $a_1, a_2, b_1, b_2, c_1, c_2, c_3$ and c_4, together with eight aliquots from a common reference sample, $r_i = r$, $i = 1, 2, 3, \cdots, 8$, are prepared for the experiment. Samples a_1 and a_2 belong to one condition (or phenotype, class); samples b_1 and b_2 to the other condition; and samples c_1, c_2, c_3 and c_4 to yet another condition. For example, a_1 and a_2 may be the genomic DNA from the blood of two schizophrenia patients, b_1 and b_2 the genomic DNA from the blood of two bipolar disorder patients, and c_1, c_2, c_3 and c_4 from the blood of four healthy controls. The common reference can be from the genomic DNA of a yet independent healthy individual. The design and labeling of the experiment are shown in Figure 5.3. Notice of the dye-swapping, that is, the reference samples in microarrays $a_1 \leftarrow r$ and $a_2 \rightarrow r$ differ in the labeling. The aim of the experiment is to find sequences that show significant differential methylation across conditions.

Since there are three conditions from eight distinct biological sources, we are to estimate the three mean log ratios, each corresponding to a condition;

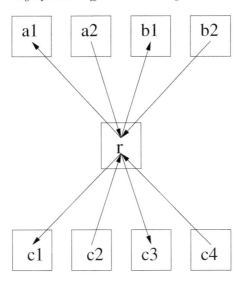

FIGURE 5.3:   A common reference design with dye-swaps. A microarray is represented by an arrow whose head represents Cy5 labeling and tail Cy3 labeling. Sample *a*s belong to a class, sample *b*s to a second class and sample *c*s to a third class.

namely, β_a, β_b and β_c, where β_a represents the mean log ratio of samples a_1, a_2 to r, and β_b the mean log ratio of samples b_1, b_2 to r, and so forth. β_a, β_b and β_c, the quantities of interest, are called coefficients (or parameters) in linear model. We will find it convenient to represent the coefficients in a column vector $\vec{\beta} = (\beta_a, \beta_b, \beta_c)^T$, where the superscript ^{T} stands for matrix transpose. Note that $(\beta_a, \beta_b, \beta_c)$ is a 1 by 3 matrix or row vector and $(\beta_a, \beta_b, \beta_c)^T$ becomes a 3 by 1 matrix or column vector. The symbols and notations should present no hurdle. They are meant to make formulas neater.

In the analysis, background corrected Cy3 and Cy5 intensities are read in. After taking log of the intensity ratios and performing the ritual of normalization (cf. chapter 4), we need to let the analysis know what coefficient a microarray's measurement contributes to the estimate of. In the example of

Figure 5.3, the relations are

$$
\vec{y} =
\begin{pmatrix}
\log(a_1/r) \\
\log(r/a_2) \\
\log(b_1/r) \\
\log(r/b_2) \\
\log(c_1/r) \\
\log(r/c_2) \\
\log(c_3/r) \\
\log(r/c_4)
\end{pmatrix}
$$

$$
=
\begin{pmatrix}
1 & 0 & 0 \\
-1 & 0 & 0 \\
0 & 1 & 0 \\
0 & -1 & 0 \\
0 & 0 & 1 \\
0 & 0 & -1 \\
0 & 0 & 1 \\
0 & 0 & -1
\end{pmatrix}
\begin{pmatrix}
\beta_a \\
\beta_b \\
\beta_c
\end{pmatrix}
+
\begin{pmatrix}
N(0,\sigma^2) \\
N(0,\sigma^2) \\
N(0,\sigma^2) \\
N(0,\sigma^2) \\
N(0,\sigma^2) \\
N(0,\sigma^2) \\
N(0,\sigma^2) \\
N(0,\sigma^2)
\end{pmatrix}
\tag{5.1}
$$

$$
= X\vec{\beta} + \vec{\varepsilon},
$$

where the vector $\vec{y}$ holds the data, the 8×3 matrix, X, is called design matrix, and $N(0,\sigma^2)$ is a draw from the normal distribution of mean zero and standard deviation σ. That is, the errors or fluctuations on the spot from the eight microarrays are independent and identically distributed (i.i.d.) random variables. It is also understood we have an equation like (5.1) for every spot on the microarray and that the same analysis is independently done for every spot. In other words, we have assumed noncorrelation between the sequences. We also dropped the index on spot in equation (5.1) for simplicity.

Once a model like equation (5.1) is written down, the rest is straightforward. For example, to get an estimate of the unknown coefficients β_a, β_b and β_c, trial values of $\beta'_a, \beta'_b, \beta'_c$ are iteratively adjusted so that the sum of squared residuals, SS_R,

$$
\begin{aligned}
SS_R = {} & [\log(a_1/r) - \beta'_a]^2 + [-\log(r/a_2) - \beta'_a]^2 + [\log(b_1/r) - \beta'_b]^2 + \\
& [-\log(r/b_2) - \beta'_b]^2 + [\log(c_1/r) - \beta'_c]^2 + [-\log(r/c_2) - \beta'_c]^2 + \\
& [\log(c_3/r) - \beta'_c]^2 + [-\log(r/c_4) - \beta'_c]^2
\end{aligned}
\tag{5.2}
$$

is minimized subject to constraints if any. In fact, since the model is linear, analytical solutions exist for the best estimate of the coefficients $\vec{\beta}$,

$$
\hat{\vec{\beta}} = (X^T X)^{-1} X^T \vec{y},
\tag{5.3}
$$

where the hat ˆ stands for estimate. An independent estimate for the error variance σ^2 in equation (5.1) is the sum of squared residuals divided by the degrees of freedom,

$$\hat{\sigma}^2 = \frac{SS_R}{N-p} = \frac{(\vec{y} - X\hat{\vec{\beta}})^T (\vec{y} - X\hat{\vec{\beta}})}{N-p} , \qquad (5.4)$$

where N is the number of microarrays and p the number of coefficients. Their values are $N = 8$ and $p = 3$ for the example of equation (5.1).

Because of one channel in oligonucleotide chips, we use log intensities, instead of log ratios, for the $\vec{y}$ in the linear model equation for single-color oligonucleotide chips. Likewise, there will be no dye-swapping oligonucleotide chips and the -1s in the design matrix X will all be 1s for experiments with single-color oligonucleotide chips.

5.2.2   Sequence-specific dye effect in two-color microarrays

Sequence-specific dye effect occurs when one of the dyes binds better to the target sequence in the labeling reaction. The consequence is that the relative intensities from the spot is shifted by a value, which is specific to the probe sequence. The value, or magnitude of the bias, reveals itself when the true log intensity ratio is zero. In the language of linear regression, the bias is an intercept term on the right-hand side of equation (5.1).

True log intensity ratios are unknown to us. They are actually the quantities we are going to find out through the experiment. Moreover, they are not necessarily zeros. The bias, if any, can be made evident with dye-swapping duplicate measurements. Suppose the log ratio reads 3 in one microarray while it reads -1 in the other duplicate whose dye orientation is reversed with respect to the former. The average of the log ratios is $[3 - (-1)]/2 = 2$. The bias, therefore, is $3 - 2 = 1$. We see that with an offset value of 1, the log ratios from the pair of dye-swapping measurements become balanced about the offset. The dye effect thus produces a nonvanishing baseline level in the ratios.

To take into account sequence-specific dye bias using dye-swapping microarrays, we expand the coefficient vector $\vec{\beta}$ to include a coefficient β_{dye} for dye effect, i.e., $\vec{\beta} = (\beta_{dye}, \beta_a, \beta_b, \beta_c)^T$ in the example of equation (5.1). The design matrix X will have to be expanded accordingly to instate the coefficient β_{dye},

$$\vec{y} = X\vec{\beta} + \vec{\varepsilon} = \begin{pmatrix} 1 & 1 & 0 & 0 \\ 1 & -1 & 0 & 0 \\ 1 & 0 & 1 & 0 \\ 1 & 0 & -1 & 0 \\ 1 & 0 & 0 & 1 \\ 1 & 0 & 0 & -1 \\ 1 & 0 & 0 & 1 \\ 1 & 0 & 0 & -1 \end{pmatrix} \begin{pmatrix} \beta_{dye} \\ \beta_a \\ \beta_b \\ \beta_c \end{pmatrix} + \vec{\varepsilon} . \qquad (5.5)$$

With the added column of 1s in X, the coefficient for dye bias appears as an intercept term in the linear model. Appendix of this chapter presents more design examples and their linear models. Since different experimental designs end up with different design matrices, familiarity with design matrix formulation helps straighten up subsequent data analysis.

5.3   *t*-test for contrasts

Equation (5.3) finds the best estimates for the log-ratios at differing conditions from the intensity data. We can now form contrasts of our interest. For example, we are interested in the comparison of methylation status between conditions a and c. Since $\hat{\beta}_a$ is an estimate of $\log(a/r)$ and $\hat{\beta}_c$ an estimate of $\log(c/r)$, we form a contrast between $\hat{\beta}_a$ and $\hat{\beta}_c$ by $\hat{\beta}_a - \hat{\beta}_c$, which represents a best estimate of the difference $\beta_a - \beta_c = \log(a/r) - \log(c/r) = \log(a/c)$, i.e., the log fold change between the two conditions a and c. Since there can be so many different constructs of contrasts, we, again, introduce contrast matrix $\vec{c}$ to simplify notations. In this example, with $\vec{c}^T = (0, 1, 0, -1)$,

$$\vec{c}^T \hat{\vec{\beta}} = (0, 1, 0, -1) \begin{pmatrix} \hat{\beta}_{dye} \\ \hat{\beta}_a \\ \hat{\beta}_b \\ \hat{\beta}_c \end{pmatrix} = \hat{\beta}_a - \hat{\beta}_c , \qquad (5.6)$$

is the contrast of our interest. Similarly, $\vec{c}^T = (0, 0, 1, -1)$ will compare condition b with condition c and $\vec{c}^T = (0, 1, -1, 0)$ enables a contrast between conditions a and b. If our interest is in contrasts between a condition and the reference, the contrast matrix can be $\vec{c}^T = (0, 1, 0, 0)$, $(0,0,1,0)$ or $(0,0,0,1)$. If we are to seek any dye effect, the contrast matrix will be $\vec{c}^T = (1, 0, 0, 0)$.

Linear models do not pay off if they do not go beyond fold change. Because of the error εs in the model (5.1) or (5.5), estimation of contrast $\vec{c}^T \hat{\vec{\beta}}$ is subject to uncertainty (see also equation (5.3)). Once we know how the estimates distribute in general, we can assess the significance of any particular contrast between the estimates. Statistically, we put forward a null hypothesis of no difference in the contrast H_o and an alternative hypothesis H_a that is the negative of H_o,

$$\begin{aligned} H_o &: \vec{c}^T \hat{\vec{\beta}} = 0 \quad \text{e.g. } \log(\hat{a}/c) = 0 \\ H_a &: \vec{c}^T \hat{\vec{\beta}} \neq 0 \quad \text{i.e. } \log(\hat{a}/c) \neq 0 . \end{aligned} \qquad (5.7)$$

We then test the null hypothesis by calculating the *t*-statistic, which is the estimated effect size, $\vec{c}^T\hat{\vec{\beta}}$, in units of its estimated standard deviation, $\hat{\text{std}}(\vec{c}^T\hat{\vec{\beta}})$,

$$t\text{-statistic} = \vec{c}^T\hat{\vec{\beta}}/\hat{\text{std}}(\vec{c}^T\hat{\vec{\beta}})$$

$$= \vec{c}^T\hat{\vec{\beta}}/\sqrt{\hat{\sigma}^2\vec{c}^T(X^TX)^{-1}\vec{c}} \qquad (5.8)$$

$$\sim t_{N-p}\,.$$

The t-statistic can be shown to follow a Student's t-distribution of $(N-p)$ degrees of freedom. We can then let computers find the corresponding p-value of the t-statistic. For example, in R (cf. chapter 13), the two-sided p-value is obtained via the command, `2*(1-pt(abs(t-statistic),N-p))`, where `pt()` and `abs()` are built-in functions in R, returning, respectively, the probability and absolute value of the input. The manual of the function can be retrieved by the command `help(pt)` in the R environment.

A p-value is the probability that a probe's DNA methylation (or gene's expression) are different between the two conditions due to chance. Specifically, it is the probability value of getting the test statistic at least as extreme as that would be obtained by chance alone, given that the null hypothesis H_o is true. The smaller the p-value, the less likely that chance is the only operator behind the observation granted by H_o. That is, a small p-value provides evidence against the null hypothesis. Statisticians typically use 0.05 as the significance level, a p-value below which spells that the alternative hypothesis, i.e., fold change in our case, is significant. With the linear models (5.1) or (5.5) and t-statistic (5.8), we savor the flavor of statistical significance.

5.4   *F*-test for joint contrasts

In larger experiments, we very often want to test multiple contrasts simultaneously (i.e., multiple testing). For example, we want to pick up differentially methylated loci between conditions a and c and between conditions b and c. Likewise, in a time course methylation experiment, we are interested in identifying loci that are differentially methylated between any successive time points. We then stack up individual contrasts to form an augmented contrast matrix, C,

$$C\hat{\vec{\beta}} = \begin{pmatrix} 0 & 1 & -1 & 0 \\ 0 & 1 & 0 & -1 \\ 0 & 1 & 0 & 0 \end{pmatrix} \begin{pmatrix} \hat{\beta}_{dye} \\ \hat{\beta}_a \\ \hat{\beta}_b \\ \hat{\beta}_c \end{pmatrix} = \begin{pmatrix} \hat{\beta}_a - \hat{\beta}_b \\ \hat{\beta}_a - \hat{\beta}_c \\ \hat{\beta}_a \end{pmatrix} = \begin{pmatrix} \log(\hat{a}/b) \\ \log(\hat{a}/c) \\ \log(\hat{a}/r) \end{pmatrix}. \qquad (5.9)$$

The null hypothesis for the multiple contrasts is

$$H_o : C\hat{\beta} = 0 \text{ i.e. } \log(\hat{a/b}) = 0, \ \log(\hat{a/c}) = 0, \ \log(\hat{a/r}) = 0$$
$$H_a : \text{at least one of } \log(\hat{a/b}) \neq 0, \ \log(\hat{a/c}) \neq 0, \ \log(\hat{a/r}) \neq 0 \ . \qquad (5.10)$$

To test the null hypothesis, we follow the technique of analysis of variance (ANOVA) by comparing the estimated standard deviation that is obtained assuming $C\hat{\beta} = 0$ (i.e., the null model) with that assuming $C\hat{\beta} \neq 0$ (i.e., the full model). The result is an F-statistic; it essentially measures the relative increase in fitting errors when we move from a full model to a null model. With SS_{R_n}, df_n, SS_{R_f} and df_f for the sums of squared residuals and degrees of freedom for the null and full models, respectively, it can be shown that a least square estimate of parameters leads to

$$F\text{--statistic} = \left[(SS_{R_n} - SS_{R_f})/(df_n - df_f) \right] / \left[SS_{R_f}/df_f \right]$$

$$= (C\hat{\beta})^T \left(C(X^T X)^{-1} C^T \right)^{-1} (C\hat{\beta})/(\hat{\sigma}^2 q) \qquad (5.11)$$

$$\sim F_{q, N-p} \ ,$$

which follows, if H_o is true, an F-distribution of q numerator and $(N - p)$ denominator degrees of freedom with q the number of (independent) contrasts in the hypothesis testing.

We get the p-value for $F_{q,N-p} > F$-statistic by calling functions in R, clicking a button in an analysis software or looking up percentile tables. The larger the F-statistic, the smaller the p-value. The p-value lets us decide whether the increase in fitting errors is large enough to reject exclusion of the parameters (i.e., with $C\hat{\beta} = 0$). p < 0.05 is normally accepted as the significance level of rejecting a null hypothesis in favor of the alternative. If the null hypothesis is rejected, it is an indication that the sequence exhibits differential methylation in at least one of the contrasts. If we are interested in the particular contrast(s) that is (are) attributable to the methylation, we then perform individual t-tests as in the last section.

t- or F-statistics and their distributions under the null hypothesis help us assess the statistical significance in the estimated fold changes between groups of samples. It is a great addendum to the threshold to fold change method in identifying differentially methylated sequences (expressed genes). Now, for every probe sequence, we have both mean fold change and p-value. We plot the fold change in one axis and $-\log_{10}$ of the p-value on the other for every probe on the microarray to form the so-called volcano plot. The larger the $-\log_{10}(\text{p-value})$, the smaller the p-value. It, therefore, serves as a measure of statistical significance of the estimated fold change. Figure 5.4 shows an example of volcano plot.

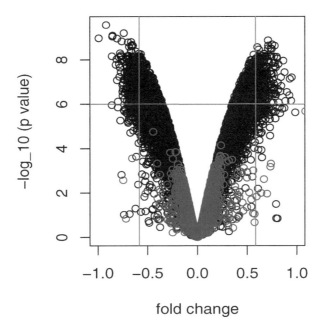

FIGURE 5.4: Volcano plot of twins versus co-twins in black and self versus self in blue. Spots outside the two vertical red lines and below the horizontal red line are possible false positives by the fold change method alone. Spots inside the two vertical lines and above the horizontal line are possible false negatives by the fold change method. Spots in the upper left and right corners are promising candidates for further investigation. The self–self hybridizations in this example help place the vertical cuts.

5.5   P-value adjustment for multiple testing

Linear models such as equation (5.1) apply to individual spots. t- or F-statistics are then calculated, spot by spot, to test for significant fold changes over the contrasts. If the p-value of a contrast is small enough, we reject the null hypothesis of no fold change and claim a positive finding (i.e., a DNA locus causing the phenotypic difference under study is predicted). Given a critical p-value, α, of 0.05, the probability of falsely rejecting a null hypothesis is 5 percent. That is, the null hypotheses are rejected by chance, not by truly differential methylation among the samples. The false rejections are called false positives. Consider a microarray consisting of ten thousand DNA sequence probes. An α of 0.05 will, on average, return five hundred false positives even though in reality none of the ten thousand sequences are differentially methylated (expressed)! The large number of erroneous findings means that huge resources will be wasted in follow-up experiments. A simple way to save waste, and to keep the researcher's reputation as well, is to lower the critical p-value α.

5.5.1   Bonferroni correction

Suppose that the rate of an auto accident per person per day is $1/10,000$. The rate is low, but still we hear accounts of auto accidents every day. The reason is that the entire population is tested in auto safety every day. The population-wise auto accident rate, therefore, is appreciable. Similarly, a high throughput microarray experiment interrogates a large number of DNA fragments in parallel. The experiment-wise error rate, i.e., the chance of any false positives in an experiment, can become appreciable even though the spot-wise error rate α is already set to a small value of 0.05.

One way of controlling the experiment-wise error rate is to lower the critical p-value by a factor of G, i.e.,

$$\alpha = \alpha/G , \qquad (5.12)$$

where G is the number of probes in a microarray and $=$ is an assignment operator. Note that in the case of multiple contrasts as in equation (5.10), G is the number of probes multiplied by the number of simultaneous contrasts. With this Bonferroni correction for critical p-value, the chance of any false positives in the experiment, $1 - (1 - \alpha/G)^G$, remains as low as α:

$$1 - (1 - \alpha/G)^G \approx \alpha . \qquad (5.13)$$

5.5.2   False discovery rate

Reducing false positives by controlling experiment-wise error rate is considered too conservative in that true positives are hard to emerge due to loss

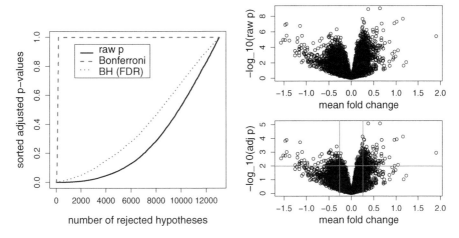

FIGURE 5.5: Multiple testing corrections (left) and volcanoes before (right top) and after (right bottom) FDR correction. The number of rejected hypotheses is the number of probes on the microarray that will survive a user prescribed cutoff (a horizontal line). Bonferroni correction amounts to multiplying the raw p-value by 12,192, the total number of tests (i.e., probes). If, after multiplication, the value is over 1, it is set to 1 because probability does not exceed 1.

of power (i.e., increased false negatives). A middle-ground (cf. Figure 5.5) between the unadjusted p-value and the stringent experiment-wise error control is to limit the so-called false discovery rate (FDR), which is the expected proportion of false positives among the claimed positives [Benjamini95],

$$\text{FDR} = \frac{\text{\# of false positives}}{\text{\# of false positives} + \text{\# of true positives}} . \tag{5.14}$$

(FDR is set to zero if the denominator is zero.) After FDR adjustment of the p-values, a cut-off at 0.3 means that we expect, on average, seven out of ten predicted positives to be true positives.

Procedures that consider the joint probability distribution of test statistics for each hypothesis have been proposed in order to account for the simultaneous test of multiple hypotheses. Algorithms have been developed to adjust p-values for FDR control [Benjamini95]. For example, in R, the function `p.adjust(raw_p_values, method="fdr")` returns p-values adjusted using the FDR method given the raw p-values in the vector `raw_p_values`. Figure 5.5 shows an example of volcano plots before and after FDR correction.

5.6   Modified t- and F-test

The classical t-test equation (5.8) for two-condition comparison and F-test equation (5.11) for multicondition comparison improve upon the fold change method for identifying differentially methylated loci. The point is that a mean fold change or, more generally, a linear combination of the estimated fold changes are associated with a random error. Given an estimate for the error in the fold change, we can determine the level of significance of the fold change between conditions.

When the sample size (i.e., number of samples per condition) is small, it is known that estimation becomes unreliable. It can, for example, happen that the estimated standard deviation $\hat{\sigma}$ of equation (5.4) is too small due to lack of samples. Since the estimated standard deviation $\hat{\sigma}$ appears in the denominator of equation (5.8) and equation (5.11), the resulting t- and F-statistic will be too large. Modifications to the classical t- or F-statistics, therefore, were proposed to regularize the statistic [Efron01, Lönnstedt02]. One can for example replace the $\hat{\sigma}$ in equation (5.8) and equation (5.11) with $(\sigma_o + \hat{\sigma})/2$ where σ_o is an empirical constant. The replacement effectively places a lower bound to the estimated standard deviation. A candidate for σ_o is the standard deviation of the log-ratios of all the spots on the microarray. The next job will be the p-value calculation since the modified statistic may not follow the classical t- or F-distributions any more. A permutation analysis can assist in evaluating the p-value.

The standard deviation of the log ratios (or log intensities) of a sequence is sequence dependent (cf. Figure 3.8). Recall that, for simplicity, we dropped the index g in the linear model including σ_g. A microarray provides simultaneous measurements of the methylation (or expression) levels of thousands of loci (genes). For a better estimation of the individual σ_gs, it would be beneficial that information on σ_g from other loci (genes) is shared. This is of particular use when the sample size in the experiment is small (e.g., less than ten). An implementation that realizes the information sharing is empirical Bayes [Smyth04, Smyth05] (note, again we drop the index g on $\hat{\sigma}$),

$$\hat{\sigma}^2 = \frac{d_0 \hat{\sigma}_0^2 + (N - p)\hat{\sigma}^2}{d_0 + (N - p)} \ , \tag{5.15}$$

where $\hat{\sigma}_0^2$ is an inverse Gamma prior on the variance and d_0 is a constant to be estimated from the data of all loci (genes). When we substitute the modified sample variance into the statistic, it can be shown that the modified t- and F-statistics follow t- and F-distributions under the null hypothesis with $(d_0 + N - p)$ and $(q, d_0 + N - p)$ degrees of freedom [Smyth04]. The extra degrees of freedom d_0 come from the fact that information is borrowed from the ensemble of loci (genes) for statistical inference about individual loci (genes).

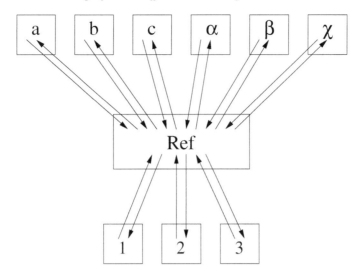

FIGURE 5.6: A reference design with dye-swaps for within-group and between-group variation study. Samples a, b, c belong to a group; samples α, β and χ to a second group; and samples 1, 2, and 3 to a third group. A microarray is represented by an arrow whose head represents Cy5-labeled sample and tail Cy3-labeled sample.

5.7 Significant variation within and between groups

Loci whose methylation are highly variable (or stable) between individuals within and among populations may suggest their importance in clinics as well as in evolution. As an illustration, we apply linear models to address the issue of significant variations in methylation within and between groups. Figure 5.6 shows an experimental design comparing variations between the three samples within each group and among three groups using eighteen microarrays.

5.7.1 Within-group variation

We begin with the identification of loci that show significant variation in methylation between samples within the same groups. The null hypothesis states that there exist no differences between the samples within each group, while the alternative hypothesis says the opposite. That is,

$$H_o: \quad \mu_a = \mu_b = \mu_c, \ \mu_\alpha = \mu_\beta = \mu_\chi, \ \mu_1 = \mu_2 = \mu_3$$
$$\text{and } \mu_a \neq \mu_\alpha \neq \mu_1 \qquad (5.16)$$

$$H_a: \mu_a \neq \mu_b \neq \mu_c \neq \mu_\alpha \neq \mu_\beta \neq \mu_\chi \neq \mu_1 \neq \mu_2 \neq \mu_3.$$

The full model for the alternative hypothesis has nine coefficients corresponding to the total of nine samples in the three groups (cf. Figure 5.6). The reduced model for the null hypothesis, on the other hand, has only three coefficients corresponding to the three different groups because now the within-group samples do not differ under the null hypothesis. We fit the two linear models to the data, getting their mean sums of squared residuals. We then form the F-statistic as in equation (5.11) and look up the tabulated $F_{9-3,18-9}$-distribution for the p-value. We repeat the fitting for each locus and rank the loci according to the p-values. Loci whose methylation vary significantly within the groups can be identified given a cut-off on the p-values.

5.7.2 Between-group variation

The test of between-group variation is the canonical one, i.e., to see if any of the group means is different from another. We average each dye-swapping microarray pair to form averaged log ratios for the spot on the microarray:

$$\log_2 R/G = \frac{\log_2 R/G - \log_2 R'/G'}{2}, \qquad (5.17)$$

where $\log_2 R'/G'$ is from the dye swapping microarray. The nine averaged log ratios of a spot are then fitted to the two representations of the linear model where one respects grouping of samples and the other does not. That is,

$$H_o: \mu_{\{a,b,c\}} = \mu_{\{\alpha,\beta,\chi\}} = \mu_{\{1,2,3\}}$$
$$H_a: \text{at least two of them are different}. \qquad (5.18)$$

The mean sums of squared residuals returned from the regressions are then used to form the F-statistic. The p-value of a spot is obtained from the $F_{3-1,9-3}$-distribution.

As the example in Figure 5.7 shows, with the proportions of significantly variable loci identified within groups and between groups, we are able to quantify the difference between the within-group and between-group variations. Figure 5.8 is a "volcano" plot of the multisample F-test that is equivalent to Figure 5.4 for two-sample t-test. Because of the diversity in gene functions, we might focus on a subset of genes on the microarray that pertain to a specialized biological pathway for the variation study. Similarly, in DNA methylation, we focus on the loci on few chromosomes. We may also need to justify the

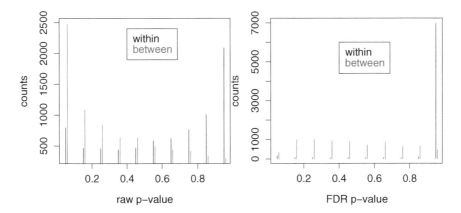

FIGURE 5.7:   Raw- (left) and FDR-adjusted (right) p-values of the F-test for sample variation within- and between-groups. Note that the red bars are shifted right a bit for display purpose. Data were from a design similar to Figure 5.6, but of a larger scale.

way the microarrays are normalized between arrays because, for example, few outlier microarrays might dominate the between-group variation.

Note also that if the group means are close to one another, we might need to have an enough number of groups or samples per group in order to reject the null hypothesis of no variation in means at a certain confidence level with, say, 90 percent chance. For example, given a within-group standard deviation equal to 0.25, we need *seven* independent samples per group in order to detect a 1.6-fold difference in one of the three groups (corresponding to a between-group standard deviation of $0.4 = \text{std}(0, 0, \log_2 1.6)$) with a power of 0.9 and false positive rate of 0.001. The sample size issue for ANOVA test here is similar to that for two-sample t-tests encountered in section 3.7.2 during experimental design.

5.8   Significant correlation with a co-variate

We might be interested in identifying loci whose methylation levels increase (or decrease) with the levels of a factor, such as age, lifetime antipsychotics or any other quantitative traits. In these cases, we profile the methylation of, say, twenty individuals, each with a pair of dye-swapping microarrays in a common reference design. To correct for the dye bias, we again find the mean methylation ratios from the pair of dye-swapping technical duplicates by equation (5.17). We then calculate the Pearson product-moment correlation

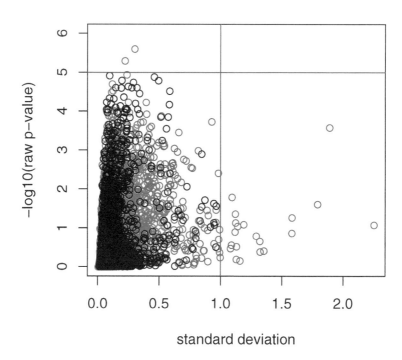

FIGURE 5.8: Volcano plot of the *F*-test results of Figure 5.7. Black and red circles are respectively the within- and between-group standard deviations of the fold changes under the settings of equation (5.18). Loci in the upper right region exhibit significant fold changes between conditions and are promising candidates for further studies. None appear in the present example.

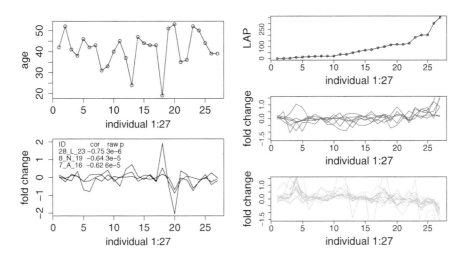

FIGURE 5.9: Loci whose methylation status correlates with age in years (left) and lifetime antipsychotics (right) in arbitrary units.

coefficient (cf. section 1.1.3.1) between the twenty mean log ratios and the twenty ages for each locus on the microarray. We can rank the loci according to the magnitudes of the correlation coefficients. However, we need a measure of statistical significance, i.e., p-value, otherwise an arbitrary threshold to the correlations would be subjective.

To get the p-value of the correlation coefficient, we randomly permute the ages of the individuals. We then calculate the Pearson correlation coefficient between the twenty log ratios and the twenty permuted ages, for each locus. We repeat the permutation procedure ten thousand times and get a distribution of ten thousand artificial Pearson correlation coefficients for each locus. The distribution is what the correlation coefficients would distribute under the null hypothesis of no correlation between the methylation and age. Let's look at the first locus first. Assume its real correlation coefficient is 0.6. If among the ten thousand artificial correlation coefficients of this locus, only ten are larger than 0.6, then the one-sided p-value of this correlation coefficient is $10/10,000 = 0.001$. If the correlation coefficient of the second locus is -0.3 and among the ten thousand artificial correlation coefficients for this locus, five thousand are less than it, then its p-value is $5,000/10,000 = 0.5$. We continue the evaluation to get the p-values of the rest of the loci on the microarray. We next correct the raw p-values for multiple testing by FDR. We can plot the correlation coefficient on the x-axis and $-\log_{10}$ of the FDR-adjusted p-values on the y-axis. We will get a volcano plot similar to that of Figure 5.4. Figure 5.9 shows examples of the loci that are identified to correlate with the age and lifetime antipsychotics.

5.9   Permutation test for bisulfite sequence data

We have so far focused on methods for identifying differential methylation at prescribed levels of significance. In short, with linear models on log transformed intensities, error terms due (mainly) to biological variability are estimated. The error inevitably propagates into the estimate of contrasts of biological interest. Knowing the distribution of the estimates, we are able to assess the statistical significance of the differential methylation. We then rank the differentially methylated loci in order of their significance.

CpG island microarrays, together with methylation-sensitive restriction enzymes, are a high throughput technology for DNA methylation profiling. The importance of DNA methylation lies in the finding that, in mammals, the degree of DNA methylation in the regulatory regions often inversely correlates with the expression of corresponding genes. Outliers in volcano plots, such as Figure 5.4 and Figure 5.5 identify candidate loci that show differential methylation among samples. After some bioinformatic research, such as to remove those loci with SNPs and copy number variations, we select a few promising loci for confirmation by an independent technique, such as sequencing of bisulfite-treated DNA. The resulting data are a string of 0s and 1s for the CpGs in a sequence clone (e.g., 1 corresponds to a methylated cytosine and 0 to an unmethylated cytosine). The question of interest remains the same: Is the methylation profile in one DNA sample (or group of samples) different from that in another DNA sample (or group of samples)? In this case of bisulfite sequencing, the data presented to the analyst are of dramatically different nature. Linear models (5.1) for log ratios or log intensities are not applicable here.

With the bisulfite sequencing method, a sample's methylation profile is represented by a collection of sequenced clones. Suppose the DNA fragment under investigation contains thirty CpG sites. The thirty 0s and 1s are stored in a row vector per clone. So, for example, in the case of six clones per sample,

$$
\vec{A} = \begin{pmatrix}
0\,0\,1\,0\,0\,1\,0\,1\,0\,1\,1\,1\,0\,0\,0\,1\,0\,0\,0\,1\,0\,1\,1\,1\,0\,0\,0\,0\,0\,1 \\
0\,0\,0\,1\,0\,1\,0\,1\,0\,0\,0\,0\,1\,1\,1\,1\,0\,0\,0\,1\,0\,1\,0\,0\,0\,1\,0\,1\,0\,1\,0\,1 \\
1\,1\,1\,1\,0\,0\,1\,0\,0\,0\,0\,1\,0\,1\,0\,0\,0\,0\,1\,0\,0\,1\,0\,1\,0\,1\,1\,1\,1\,1 \\
0\,0\,0\,0\,0\,1\,0\,1\,0\,0\,0\,1\,0\,1\,0\,0\,0\,0\,1\,0\,0\,1\,0\,1\,1\,0\,1\,0\,1\,0 \\
1\,0\,1\,1\,1\,1\,0\,1\,0\,0\,1\,0\,0\,0\,0\,1\,0\,0\,1\,0\,0\,1\,0\,0\,0\,0\,1\,1\,0\,1 \\
0\,0\,1\,0\,0\,1\,0\,1\,0\,0\,1\,1\,0\,1\,1\,1\,0\,0\,1\,0\,1\,1\,0\,0\,0\,0\,1\,1\,0\,1
\end{pmatrix}
$$

$$(5.19)$$

$$
\vec{B} = \begin{pmatrix}
0\,0\,0\,1\,0\,1\,0\,1\,0\,0\,1\,1\,0\,0\,0\,1\,0\,1\,0\,1\,1\,0\,0\,1\,0\,1\,0\,1\,1\,0 \\
0\,0\,1\,1\,0\,1\,1\,0\,0\,1\,1\,0\,0\,1\,1\,1\,1\,0\,1\,1\,0\,1\,0\,0\,0\,0\,1\,0\,0\,1 \\
0\,1\,1\,0\,0\,1\,0\,1\,0\,0\,1\,1\,0\,1\,0\,1\,0\,0\,0\,1\,0\,1\,0\,1\,1\,1\,1\,0\,0\,1 \\
0\,0\,0\,1\,1\,0\,0\,1\,0\,1\,1\,1\,0\,1\,0\,0\,0\,0\,1\,0\,0\,1\,0\,1\,0\,0\,1\,1\,1\,1 \\
0\,0\,1\,1\,0\,1\,0\,0\,1\,0\,1\,0\,0\,1\,0\,0\,1\,0\,1\,1\,0\,0\,0\,0\,0\,0\,0\,1\,0\,1 \\
0\,0\,1\,1\,0\,0\,1\,0\,1\,0\,1\,1\,1\,0\,0\,1\,0\,0\,1\,1\,0\,1\,1\,1\,0\,1\,1\,1\,0\,0
\end{pmatrix},
$$

represent, respectively, samples A and B. The task is then to determine if the methylation profile at the locus represented by matrix $\vec{A}$ is the same as that by matrix $\vec{B}$. Note that the order of the rows in the matrix is irrelevant as we can put a clone in any row.

5.9.1 Euclidean distance

First of all, we need to define a metric based on which dissimilarity can be quantified. The common one is Euclidean distance, which is simply the ordinary three-dimensional distance extended to any higher dimensions. Next, in analogy to the probeset for a gene in a single-color oligonucleotide chip, we summarize the matrix into a single row vector by averaging along the column. So, for example, the first element in the summarized row vector $\bar{A}$ is obtained from the first column of matrix $\vec{A}$ by $\bar{A}_1 = (0 + 0 + 1 + 0 + 1 + 0)/6 = 0.33$, and similarly for the other elements in the vector. Figure 5.10 plots such summarized vectors from a real dataset. The distance d between the summarized vectors $\bar{A}$ and $\bar{B}$ is calculated,

$$d = \sqrt{\sum_{i=1}^{30}(\bar{A}_i - \bar{B}_i)^2} \, . \tag{5.20}$$

We got a distance, which alone tells little. The next key ingredient will be a measure of significance. That is to find the p-value of the distance d, which is the probability of observing a value of distance as large as or larger than d under the null hypothesis that matrix $\vec{A}$ and matrix $\vec{B}$ are obtained by sampling from the same population of clones. If the one-sided p-value is less than α, we say that we reject the null hypothesis at the significance level of α.

To get the p-value under the null hypothesis, we combine together the six rows in each of the two matrices into a pool with twelve rows. We then randomly select six rows from the pool to form an artificial matrix $\vec{A}'$. The sampling is done without replacement. The remaining six rows in the pool then form matrix $\vec{B}'$. We summarize matrices to form row vectors $\bar{A}'$ and $\bar{B}'$ as before. We then calculate the Euclidean distance d' between $\bar{A}'$ and $\bar{B}'$ according to the metric equation (5.20). We repeat the permutation procedure one thousand times. Since the selection from the pool is random, we will get one thousand different d''s.

With the frequency distribution of the distances d''s under the null hypothesis as in Figure 5.10, we find the area under the distribution curve between the rightmost end of the curve to d. The area is divided by the total area, which is one thousand in this case, to get the fractional area. The one-sided p-value of d is then the fractional area. Finally, the p-value is compared to the desired significance level α to see if we can reject the null hypothesis of no difference between the two samples.

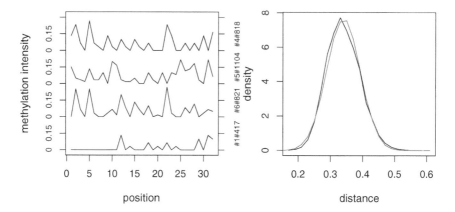

FIGURE 5.10:   Plot of the summarized vectors (left) and distribution of the distances calculated from summarized vectors from permutation (right). The goal is to test if the patterns in the left panel differ from one another. The black curve on the right is from ten thousand permutations, while the red curve is a Gaussian fit to it.

Note that since we have only twelve rows in the pool, we can have at most $C_6^{12} = 12!/(6!6!) = 924$ d's instead of one thousand mentioned above. To get a precise p-value, the distribution of d's has to be smooth, calling for a large number of d's. The number of six samples per condition, therefore, is considered a minimum for a permutation test. The method can be extended to find p-values of multicondition tests as we extend from the two-sample t-test to multisample F-test.

5.9.2   Entropy

We found the dissimilarity metric in equation (5.20) worked well in most situations. It is very sensitive to the signal profiles in the locus. For example, if the data are such that the summarized vector $\bar{A}$ contains a tall peak at site five while $\bar{B}$ a short or no peak at the same site, the Euclidean distance between the two samples will be large. That is, the two samples represented by matrices $\vec{A}$ and $\vec{B}$ are quite dissimilar under the metric. In other situations where each matrix contains a peak, but at different site inside the locus, the distance will also be large according to the definition in equation (5.20).

The Euclidean distance becomes unsuitable if biology tells us that peak locations are not important as long as there are peaks in the locus. What we need in this case is a metric that detects and counts features in the locus. Entropy, which plays a central role in information science and statistical physics, measures degree of randomness and looks useful in the present setting. If the methylation intensities are weak and each of the thirty sites in the locus (5.19) has an equal chance of methylation (i.e., the 1s in matrix $\vec{A}$ in equation (5.19) are sparse and equally distributed in it), which of the site will be methylated, if we are given a seventh clone, is least predictable. We say that the entropy of such a locus is maximal. On the other hand, if a certain site in the locus is always methylated (i.e., a column in the matrix is full of 1s), then we are willing to bet that the site will also be methylated in a new clone. The locus in this case should have a minimal entropy. A definition of entropy S that has the properties is

$$S \ = \ <\log_2(1/p)> \ = \ -\sum_{i=1}^{N} p_i \log_2 p_i \ , \qquad (5.21)$$

where the square brackets mean average, p_i is the probability of site i being methylated and N is the number of sites in the locus. Note that p_is, being probabilities, should sum up to one: $\sum_{i=1}^{N} p_i = 1$. It can be shown that the maximum entropy has a value of $\log_2 N$. We can divide equation (5.21) by this maximum value so that the normalized entropy is bounded by 1 independent of the number of sites N.

After adopting equation (5.21) as a metric of randomness in methylation at a locus, we can calculate the "distance" $d = |S_A - S_B|$ between samples A and B. We then proceed to assess the significance of difference by permutation as before.

5.10   Missing data values

Missing values are not uncommon in microarray data. These can happen when, for example, after subtracting background from foreground, the fluorescent intensity of a spot becomes zero or negative. When we take log of a nonpositive number, the value returned is undefined. Therefore, the analysis program has to be prepared for these exceptions. (Note that there exist algorithms that guarantee that all intensities are positive after background correction.)

The ways missing values are handled will depend on where they occur. If undefined values, in the form of NaN, $\pm$Inf or NA in most programming languages, enter into t- or F-tests, they can simply be removed in calculating the mean and standard deviation with the number of samples, and thus degrees of freedom, reduced accordingly.

In the case of correlation between two vectors, the counterpart elements in the other vector will have to be removed as well. So, for example, the Pearson correlation between $(1, \text{NA}, 3, 4, 5)$ and $(6, 7, 8, 9, 10)$ will be calculated between $(1, 3, 4, 5)$ and $(6, 8, 9, 10)$.

TABLE 5.1: Description of Data Files in a Factorial Design

Chip #	Wild Type (W) or Mutant (M)	Treated (T) or Untreated (U)
1	W	U
2	W	T
3	W	T
4	M	U
5	M	T
6	M	T

In Euclidean distance calculation, the sum of squared differences can be inflated in proportion to compensate for the squares that are missing due to undefined values. So, for example, the distance between $(1, 2, 3)$ and $(2, 3, \texttt{NA})$ is $\sqrt{(1^2 + 1^2) \times (3/2)} = 1.732051$.

In the entropy example, a missing value in the matrix of equation (5.19) can be replaced with 0.5, instead of removing the whole column. The imputation is fair because, assuming independence, the value that is missing can be either 0 or 1 with equal chance.

5.11   Appendix

Design matrix translates a designed experiment into linear algebra. It is a gateway between laboratory experiment and data analysis. A design matrix has its number of rows equal to the number of microarrays (or oligonucleotide chips) in the experiment. The number of columns of the design matrix is equal to the number of *independent* coefficients in the linear model. A contrast matrix maps the coefficients onto the contrasts of biological interest. This section aims to help readers familiarize themselves with the design and the associated matrices.

5.11.1   Factorial design

Table 5.1 describes a 2×2 factorial design where all the four possible different conditions are measured with six independent samples on six single-color oligonucleotide chips. Similar measurements can be accomplished by two-color DNA microarrays with a common reference. We represent the four independent coefficients in the linear model by WU, WT, MU and MT. The 6×4 design matrix that links the intensity measurements from the chips to

TABLE 5.2: Description of Data Files in a Time-Course Experimental Design

Array #	Wild Type (W) or Mutant (M)	Hours (H)	Dye Swap
1	W	0	-
2	W	0	yes
3	W	6	-
4	W	24	-
5	M	0	-
6	M	0	yes
7	M	6	-
8	M	24	-

the coefficients is then,

$$
\begin{pmatrix}
\text{log intensity in chip 1} \\
\text{log intensity in chip 2} \\
\text{log intensity in chip 3} \\
\text{log intensity in chip 4} \\
\text{log intensity in chip 5} \\
\text{log intensity in chip 6}
\end{pmatrix}
=
\begin{pmatrix}
1 & 0 & 0 & 0 \\
0 & 1 & 0 & 0 \\
0 & 1 & 0 & 0 \\
0 & 0 & 1 & 0 \\
0 & 0 & 0 & 1 \\
0 & 0 & 0 & 1
\end{pmatrix}
\begin{pmatrix}
\text{WU} \\
\text{WT} \\
\text{MU} \\
\text{MT}
\end{pmatrix} .
\tag{5.22}
$$

If we are interested in the contrast between wild/mutant treated and wild/mutant untreated and in the loci that respond differently to the treatment between strains, the contrast matrix that makes up the contrasts of interest from the coefficients will be

$$
\begin{pmatrix}
\text{WT v WU} \\
\text{MT v MU} \\
\text{(MT v MU) v (WT v WU)}
\end{pmatrix}
=
\begin{pmatrix}
-1 & 1 & 0 & 0 \\
0 & 0 & -1 & 1 \\
1 & -1 & -1 & 1
\end{pmatrix}
\begin{pmatrix}
\text{WU} \\
\text{WT} \\
\text{MU} \\
\text{MT}
\end{pmatrix} .
\tag{5.23}
$$

5.11.2 Time-course experiments

The factorial design can be easily extended to factors having more than two levels. For example, instead of treatment or none, consider the measurement of methylation difference at time in hours after the treatment in a time course experiment using two-color microarrays with reference design as outlined in Table 5.2. Note that because of two colors, allowance can be made for dye swapping as in the table.

A set of coefficients that can be modeled is W0, W6, W24, M0, M6 and M24. The design matrix has dimensions 8×6 and relates the intensity ratios from the eight microarrays to the six coefficients,

$$
\begin{pmatrix}
\text{log ratio in array 1} \\
\text{log ratio in array 2} \\
\text{log ratio in array 3} \\
\text{log ratio in array 4} \\
\text{log ratio in array 5} \\
\text{log ratio in array 6} \\
\text{log ratio in array 7} \\
\text{log ratio in array 8}
\end{pmatrix}
=
\begin{pmatrix}
1 & 0 & 0 & 0 & 0 & 0 \\
-1 & 0 & 0 & 0 & 0 & 0 \\
0 & 1 & 0 & 0 & 0 & 0 \\
0 & 0 & 1 & 0 & 0 & 0 \\
0 & 0 & 0 & 1 & 0 & 0 \\
0 & 0 & 0 & -1 & 0 & 0 \\
0 & 0 & 0 & 0 & 1 & 0 \\
0 & 0 & 0 & 0 & 0 & 1
\end{pmatrix}
\begin{pmatrix}
\text{W0} \\
\text{W6} \\
\text{W24} \\
\text{M0} \\
\text{M6} \\
\text{M24}
\end{pmatrix} .
\tag{5.24}
$$

The loci that respond to either six hours or twenty-four hours in either wild type or mutant can be sought for via the multiple contrasts,

$$
\begin{pmatrix}
\text{W6 v W0} \\
\text{W24 v W0} \\
\text{M6 v M0} \\
\text{M24 v M0}
\end{pmatrix}
=
\begin{pmatrix}
-1 & 1 & 0 & 0 & 0 & 0 \\
-1 & 0 & 1 & 0 & 0 & 0 \\
0 & 0 & 0 & -1 & 1 & 0 \\
0 & 0 & 0 & -1 & 0 & 1
\end{pmatrix}
\begin{pmatrix}
\text{W0} \\
\text{W6} \\
\text{W24} \\
\text{M0} \\
\text{M6} \\
\text{M24}
\end{pmatrix} .
\tag{5.25}
$$

Note that once (W6 v W0) and (W24 v W0) are determined, (W24 v W6) is known. Thus, there is no need for a (W24 v W6) above.

The loci that respond differently in mutant relative to wild type are identified through the contrasts,

$$
\begin{pmatrix}
\text{(M6 v M0) v (W6 v W0)} \\
\text{(M24 v M0) v (W24 v W0)}
\end{pmatrix}
=
\begin{pmatrix}
1 & -1 & 0 & -1 & 1 & 0 \\
1 & 0 & -1 & -1 & 0 & 1
\end{pmatrix}
\begin{pmatrix}
\text{W0} \\
\text{W6} \\
\text{W24} \\
\text{M0} \\
\text{M6} \\
\text{M24}
\end{pmatrix} .
\tag{5.26}
$$

5.11.3 Balanced block design

Balanced block design compares samples directly by labeling samples of one class in one channel and samples of another class in the other channel on two-color microarrays. Table 5.3 shows an example of balanced block design comparing samples of three classes. Although there are three species, the number of independent pairwise comparisons is only two, say species 1 versus 2 and species 3 versus 2. Once these two log ratios are determined, the last pairwise comparison, i.e., species 1 versus 3, can be derived. The linear model that describes the experiment of Table 5.3 involves only two coefficients. The design matrix that relates the log ratio measurements in the microarrays to

TABLE 5.3: Description of Data Files in a Balanced Incomplete Block Design

Array #	Cy3	Cy5
1	species 1	species 2
2	species 1	species 3
3	species 2	species 1
4	species 2	species 3
5	species 3	species 1
6	species 3	species 2

TABLE 5.4: Description of Data Files in a Loop Design

Array #	Cy3	Cy5
1	A	B
2	B	A
3	B	C
4	C	B
5	C	A
6	A	C

the coefficients in the linear model is

$$
\begin{pmatrix}
\log \text{Cy5/Cy3 in array 1} \\
\log \text{Cy5/Cy3 in array 2} \\
\log \text{Cy5/Cy3 in array 3} \\
\log \text{Cy5/Cy3 in array 4} \\
\log \text{Cy5/Cy3 in array 5} \\
\log \text{Cy5/Cy3 in array 6}
\end{pmatrix}
=
\begin{pmatrix}
-1 & 0 \\
-1 & 1 \\
1 & 0 \\
0 & 1 \\
1 & -1 \\
0 & -1
\end{pmatrix}
\begin{pmatrix}
\log \text{species 1/species 2} \\
\log \text{species 3/species 2}
\end{pmatrix}.
$$

$$(5.27)$$

All pairwise comparisons among the species can now be made via the contrasts,

$$
\begin{pmatrix}
\text{species 1 v species 2} \\
\text{species 3 v species 2} \\
\text{species 1 v species 3}
\end{pmatrix}
=
\begin{pmatrix}
1 & 0 \\
0 & 1 \\
1 & -1
\end{pmatrix}
\begin{pmatrix}
\log \text{species 1/species 2} \\
\log \text{species 3/species 2}
\end{pmatrix}.
\qquad (5.28)
$$

5.11.4 Loop design

Table 5.4 illustrates a loop experimental design comparing twelve samples from three classes using six two-color microarrays. Microarrays 1, 3 and 5 form a loop and microarrays 2, 4 and 6 are the dye swaps. Again, unlike the previous common reference design examples in section 5.11.1 and section 5.11.2 where the number of independent coefficients in the model is equal to the number of distinct phenotypic classes in the experiment, there are just two ratios that we can independently estimate from the six microarrays. Let the two ratios be A/B and C/B. Once these two are determined, the last, i.e., A/C, can be obtained via (A/B)/(C/B). That being said, we write down the model for this experiment,

$$
\begin{pmatrix}
\log \text{Cy5/Cy3 in array 1} \\
\log \text{Cy5/Cy3 in array 2} \\
\log \text{Cy5/Cy3 in array 3} \\
\log \text{Cy5/Cy3 in array 4} \\
\log \text{Cy5/Cy3 in array 5} \\
\log \text{Cy5/Cy3 in array 6}
\end{pmatrix}
=
\begin{pmatrix}
-1 & 0 \\
1 & 0 \\
0 & 1 \\
0 & -1 \\
1 & -1 \\
-1 & 1
\end{pmatrix}
\begin{pmatrix}
\log \text{A/B} \\
\log \text{C/B}
\end{pmatrix}.
\tag{5.29}
$$

All pairwise comparisons are made via the contrasts,

$$
\begin{pmatrix}
\text{A v B} \\
\text{C v B} \\
\text{A v C}
\end{pmatrix}
=
\begin{pmatrix}
1 & 0 \\
0 & 1 \\
1 & -1
\end{pmatrix}
\begin{pmatrix}
\log \text{A/B} \\
\log \text{C/B}
\end{pmatrix}.
\tag{5.30}
$$

Chapter 6

High-Density Genomic Tiling Arrays

A large-scale international project by the ENCODE consortium studying the regulatory elements in 1 percent of human genome revealed significant roles and activities of histone modifications in gene transcription and DNA replication. One of the major technologies in the endeavor was tiling arrays, which consist of oligonucleotide probes from the entire ENCODE region, including introns within genes and regions between genes (but excluding repetitive sequences). The pilot project also assessed and developed protocols of high-density tiling arrays and found the hybridization results were reproducible across platforms and laboratories. As a trend is seen toward tiling array assay, we dedicate this chapter to the analysis.

Common goals of DNA methylation measurements using tiling arrays include (1) identification of methylation regions in a population of samples and (2) identifying regions of differential methylation between sample populations. If there are negative control arrays, after proper intra- and interarray normalization, positive calls can be identified by nonparametric one-sample Wilcoxon test in the first application and two-sample Wilcoxon test in the second application. If no control arrays are available, in order to reduce false positive calls, we may need to set a threshold for hybridization intensity in addition to the cutoff by a critical p-value. Different probes for the same intended target can bind very different amounts of the target. When estimating a probe intensity, we require that the probes in its (genomic) proximity work together in providing a robust estimate for the probe reading. The robust estimates are related to the Wilcoxon test.

The number of tiling arrays can be refrained in the preliminary phase of an investigation. In particular, there might be no biological replicates let alone technical replicate arrays. Still, we hope to gain as much insight from the study as we can in order to guide the way toward future experiments. We introduce the technique of biplot for principal component analysis that is appropriate to the end. A biplot summarizes various methylation relations on a single two-dimensional plot.

As we upscale the wet bench technology to higher density microarrays, such as genomic tiling arrays, the number of data points (i.e., probes) increases by many hundred-fold. The demand on computer memory and speed increases as a result. Access to PC clusters in a computing facility will be desirable in undertaking such massive analysis in a timely fashion.

6.1   Normalization

6.1.1   Intra- and interarray normalization

Methods for tiling array normalization [Kampa04, Royce05, Emanuelsson06] depend on the hypothesis tests that, in turn, hinge on the experimental objectives. Some tests need no normalization. However, we describe normalizations that are applicable to most circumstances: quantile normalization for intraarray normalization and median normalization for interarray normalization.

We apply quantile normalization to array replicates on the same biological sample. Quantile normalization brings the distributions from different replicate arrays into a common location and shape. The rationale is that since the samples are from the same source, say pancreatic cells from the same individual, the methylation profiles should be more or less the same. The normalized data can be pooled together for the identification of methylation sites/boundaries in the genome.

In cases where we have samples of different origins or conditions on different arrays, we have to further normalize the data for between-array comparison. The reason is that the hybridization or scanning conditions might differ from array group to array group. A popular interarray normalization is to scale the intensities so that the median intensity of every array is the same. This is done by first dividing the intensities by the array median. We then multiply every intensity by, say, one thousand so that the median intensity of every array becomes one thousand.

6.1.2   Sequence-based probe effects

A tiling array contains millions of oligonucleotide probes that cover the one-dimensional genome at a regular interval just like tiles covering the walls of our bathroom. The length of a probe is typically twenty-five (thirty-six or sixty) bases. State of the art tiling arrays cover the nonrepetitive-sequence portion of the genome with overlaps so that the resolution (i.e., end-to-end distance between neighboring tiles) is seven bases. Before tiling microarrays, oligonucleotide arrays were developed to assay the abundance of gene transcripts, using sixteen to twenty probes for a target gene. Every 25-mer probe sequence has its optimal annealing temperature at which the target sequence binds most efficiently to the probe. However, since the hybridization is carried out at a single temperature, it is not surprising to find that probes in a probeset targeting the same transcripts register a wide range of intensities. (Note the varying intensities could be due to alternative splicing when the probes target different exons of a gene.) For example, the probes with high GC (guanine-cytosine) content in their sequences on average have higher intensities than the probes with low GC content because of the higher affinity

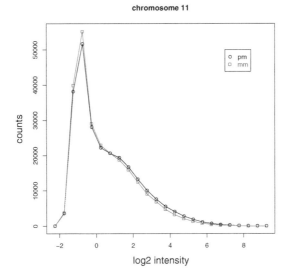

FIGURE 6.1: Distributions of the raw PM and MM intensities of a human ENCODE2 tiling array. Because of high density, only the intensities of the probes covering chromosome 11 are shown.

between G and C than between A and T. Furthermore, the effect depends not only on the content, but also on the nucleotide position in a probe. The sequence-based bias becomes worse in tiling arrays [Royce07] because, unlike the gene-centric oligonucleotide arrays where we choose sixteen to twenty optimal probes to form a probeset representing a gene, there is no room in choosing what cover the genome.

In relation to the sequence-based bias, it is found that almost all probes register intensities even though we do not expect that sequences over all the genome regions are transcribed. Unspecific hybridization, therefore, is ubiquitous in microarrays using short oligonucleotides. In Figure 6.1 we show an example of the distributions of probe intensities from a tiling array profiling DNA methylation using restriction enzyme enrichment. Indeed, we see a dominant fraction of the probes with low, yet nonvanishing, intensities even though not the entire chromosome is expected to be methylated.

The background intensity due to unintended hybridization can be estimated by control arrays to which known unmethylated (or methylated, depending on what is under profiling) samples are hybridized. The background-subtracted probe intensities are then used in downstream analysis for methylation profile or differential methylation. Since demethylated samples are not always available, an alternative is to use the promoter regions of the essential housekeeping genes as control probes. In this scheme, probes are grouped according to their GC content. The median intensity of the control probes in a GC group serves

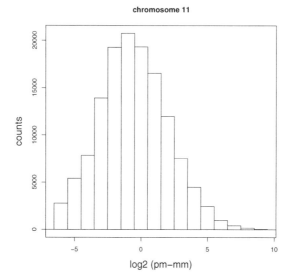

FIGURE 6.2: Distribution of $\log_2(\mathrm{PM}-\mathrm{MM})$ of the intensities in Figure 6.1. Note that since $\log_2 0$ is `-Inf` and $\log_2(\text{negative value})$ is `NaN` (not a number), the distribution shown is for the PM that are larger than MM.

as the background intensity for the other probes in the same GC group.

On the other hand, Affymetrix$^{\circledR}$ addresses the issue of unspecific hybridization by complementing each probe with a mismatch probe (MM), whose sequence is identical to the counterpart (PM), save the middle base. The idea is that fragments that bind to the mismatch probe are not specific to the perfect match probe. It was found that PM$-$MM corrected for much of the sequence based probe effect in tiling arrays although there was room for improvement. In Figure 6.2 we show the distribution of $\log_2(\mathrm{PM}-\mathrm{MM})$ from the PM and MM in Figure 6.1 with PM>MM.

6.2 Wilcoxon test in a sliding window

It was reported in the last section that, in genomic regulatory elements profiling using tiling arrays, different probes bind to the same intended targets with different affinity and that subtracting mismatch probe fluorescence is a way of lessening the effect of unspecific hybridization. The findings apply to methylation profiling because the same hybridization principle underlies microarray applications. We enriched genomic DNA with HpaII methylation sensitive restriction enzyme followed by ENCODE2 tiling array hybridization and show in Fig. 6.3 the probe (perfect and mismatch) intensities in a section of human genome. (ENCODE represents the project of ENCyclopedia Of the Dna Elements [ENCODE07], succeeding the Human Genome Project.) We see varying intensities across two hundred bases that span a typical CpG island. The variation in intensities in nearby probes can be hundreds fold as illustrated from the intensities around genomic coordinate 1712000 in Figure 6.3. Figure 6.4 shows the background (mismatch) subtracted probe intensities of the same gnomic region. Again, we see great variation in the hybridization intensities over genomic near neighbors.

We know that a robust metric for the central tendency of a non-Gaussian distribution is median. Therefore, we are tempted to utilize median as a measure of hybridization intensity that is resistant to outlying intensities. To represent the intensity at probe i, we identify the probes whose starting nucleotides are within one hundred bases upstream of the center nucleotide of probe i and the probes whose ending nucleotides are within one hundred bases downstream the center nucleotide of probe i. The window size in this case is two hundred one bases. After aggregating the readings, we calculate the median to represent the hybridization intensity of the probe defining the window. The procedure is repeated to the next probe, i.e., probe $i + 1$, $i + 2$, ..., and so on and thus is called sliding window [Kampa04].

The other argument for the windowing technique is that methylated (or transcribed) fragments are usually wider than probe length, which is typically twenty-five bases. Intensities from the neighboring probes are not independent of one another. Aggregating information from the neighboring probes in a tiling array is not only legitimate, but also advantageous.

The use of medians reminds us of the Wilcoxon tests introduced in chapter 1. To determine the significance of methylation, we resort to Wilcoxon signed-rank test for the null hypothesis of a symmetric distribution of the (PM−MM)s from the probes in a window around zero. The alternative hypothesis is a right-skewed distribution of (PM−MM)s, i.e., PM>MM. A Wilcoxon signed-rank test returns, besides p-value, a robust estimate for the PM−MM called pseudomedian [Kampa04], which is the median of the pairwise averages of the input values (PM−MM in the present case). In Figure 6.5, we show the p-value along with the pseudomedian of every probe along the genomic region of Figure 6.3 and Figure 6.4. The plot is reminiscent of the "volcano" plot showing both fold-changes and p-values from two-color microarrays in chapter 5. In Figure 6.4 is also shown the pseudomedian estimate of the PM−MM in a window of size two hundred one bases. We see that although PM−MM can swing from positive to negative between probes, pseudomedians are resilient to variations.

In general, the larger the window size, the more robust the estimation of the methylation. This is because a wider window contains more PM−MM measurements, effectively increasing the sample size. Therefore, to enhance the power of the measurement, we increase the number of replicate arrays. The PM−MM pairs from all the arrays in the same window then can be

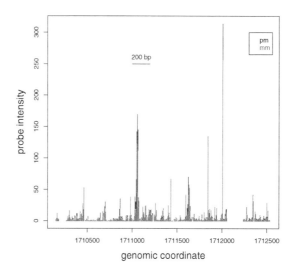

FIGURE 6.3: The PM and MM probe intensities over a region of chromosome 11 on an ENCODE2 tiling array. The gaps on the left and right represent repetitive regions and are excluded from the array design. We see that the probe intensities can change abruptly by many hundred-fold.

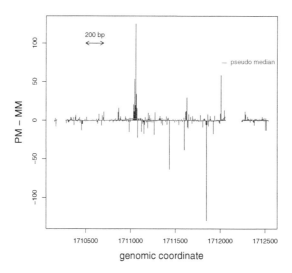

FIGURE 6.4: PM−MM of Figure 6.3 over the region of Figure 6.3. Also shown is the pseudomedian (in red) of the PM−MM in a window of size two hundred one bases. It is seen that the pseudomedian estimate of probe intensity is robust against the outlying probes in a window.

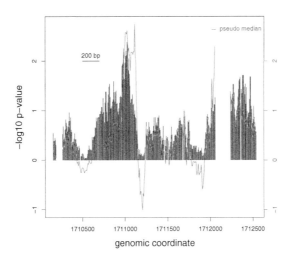

FIGURE 6.5: $-\log_{10}$(p-values) and pseudomedians of the intensities of Figure 6.4. p-values are from the Wilcoxon signed-rank tests for a symmetric distribution of PM−MM around zero. The PM−MM in a test are from the probes in a window of size two hundred one nt.

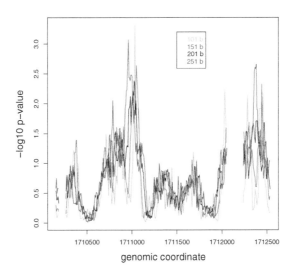

FIGURE 6.6: Wilcoxon signed-rank test p-values of the probes in a window of various sizes.

pooled together for the Wilcoxon test. Alternatively, if there is only one array available, we cannot but increasing the the window size. A drawback for wide windows is that sharp methylation boundaries are hard to detect because a large proportion of measurements are shared between neighboring windows so that estimate (pseudomedian or p-value) changes only slowly from one window to the next. In Figure 6.6, we show the Wilcoxon p-values from window sizes between 101 and 251 bases. We see that the p-value estimates start to converge for window sizes above 201.

In applications where we compare between the tiling arrays on samples at two different conditions, we pool the $\log_2(PM-MM)$ of the probes in a window from the arrays belonging to each sample group. We then run Wilcoxon rank-sum test of chapter 1 for the null hypothesis of equal location of the two distributions of $\log_2(PM-MM)$. The p-value returned from the test indicates significance of the difference in the methylation at the center of the probe, which defines the window. The test is performed from probe to probe, generating a series of p-values. In Figure 6.7, we show an example of the p-values for the methylation difference between mono- and bi-allelic expressed cells from thymus.

6.2.1 Probe score or scan statistic

So far, we have introduced two quantities for a tiling array probe, namely an estimate for the (difference in) hybridization intensity and the associated test p-value. The estimation gains robustness and power by borrowing information from nearby measurements. The situation is similar to the two-color CpG island microarray probes that have fold-changes and p-values. Question arises as to the criteria of calling a "hit." A small p-value usually signifies significant measurement; the (difference in) intensity may be large or small as long as it is consistent across the probes in the window. Therefore, we are tempted to use $-\log_{10}(\text{p-value})$ as the score for a probe or scan statistic. Whenever a probe score is above, say 5, (i.e. p-value < 0.00001), it is designated as a hit.

As was found in a recent large-scale study on cross-platform reproducibility of gene expression microarrays, hit calling based a threshold fold-change slightly improves the reproducibility [MAQC06]. To claim a hit in tiling array, we would suggest a p-value cutoff together with a threshold on the probe intensity estimate. A suitable value of the intensity threshold can be readily obtained by (a) negative control array(s).

6.2.2 False positive rate

There are as many p-values as there are probes in a high-density array. In conventional gene-centric oligonucleotide arrays, probe intensities in a probe-set are summarized to represent a gene transcript measurement. P-values in this context are associated with (differential) gene expression. If we downplay the regulation of genes in a pathway, the gene expression levels are considered

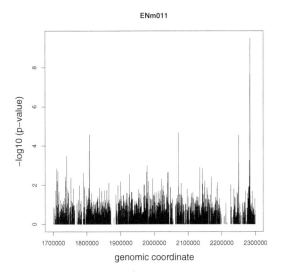

FIGURE 6.7: Wilcoxon rank-sum test p-values for the probes in 201-base windows across the ENm011 region of the ENCODE project. In a window, the PM − MM probes from the mono-allelic arrays are compared with those from the bi-allelic arrays using Wilcoxon rank-sum test.

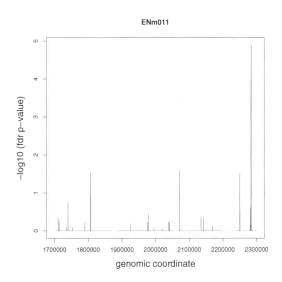

FIGURE 6.8: FDR-adjusted p-values of Figure 6.7. The regions represented by the four tallest peaks are identified as the regions that show differential methylation between the mono- and bi-allelic expressed thymic tissues.

independent of one another. We then apply multiple testing correction to the raw p-values in an effort to control the array-wise false positive rate.

The design philosophy of high-density tiling arrays is unbiased in the sense that no *a priori* weights are put on regulatory or coding sequences. We may group probes together to represent a feature, such as a gene, if we know where and what the feature is. In an exploratory study using tiling arrays, this is out of the question. Features, such as DNA methylation, can span a region wider than a probe. We, therefore, know that nearby probes are not independent. But again, before locating the feature, we are not able to group the related probes into a probeset-like entity.

The interdependence poses a problem with p-value correction for multiple testing, which usually assumes independence among the multiple tests. A way of estimating the false positive hit rate in tiling arrays is to shuffle the probes and repeat the analysis. By shuffling, say, one thousand times, a landscape of p-value distributions is obtained. We may then select a value from the null distributions as cutoff for the p-values from real data. The selected cutoff value determines the array-wise false positive rate we can tolerate. The shuffling method is conceptually easy and methodologically amenable to variant null hypotheses. It is, however, computationally intensive. An popular alternative for multiple testing correction is by the false discovery rate (FDR) introduced in section 5.5.2. In Figure 6.8, we also show the FDR corrected p-values.

6.3   Boundaries of methylation regions

One of the major purposes of using unbiased tiling arrays is to locate the methylation regions in the genome with great resolution. This is approached by array hybridization with enriched fractions from the methylated DNA immunoprecipitation technique (mDIP) in section 2.2.2. The methylation intensity pattern ideally looks like piece-wise steps (or segments) along the chromosome with varying heights from step to step. The points at which heights change are the locations of the mDIP fraction boundaries on genome.

A model to describe the behavior is by the so-called structural change model [Huber06],

$$y_i = \mu_S + \epsilon_i \ \ \text{for} \ \ S < i < S+1 \,, \tag{6.1}$$

where y_i is the intensity estimate of probe i, μ_S the height of segment S defined by the change points at probes S and $S+1$, and ϵ_i the measurement error of probe i. The index i runs over the probes, which are within the segment. Unknown parameters in the model of equation (6.1) including boundaries of the segments S and segment heights μ_S can be estimated by minimizing the sum of squared residuals $SS_R = \sum_{i,S}(y_i - \bar{y}_S)^2$, with $\bar{y}_S$ being the mean of y_i in segment S, by dynamic programming algorithms. Note that as the number of segments S_{max} increases, the sum of squared residuals decreases. Minimum SS_R occurs when the number of segments equals to the number of probes: $S_{max} = N$. A way to tax large S_{max} is by the Bayesian information criterion (BIC), which subtracts log likelihood of the fit $(2\pi SS_R/N)^{-N/2}e^{-N/2}$ by a penalty proportional to the number of parameters in the model $S \log N$ [Huber06].

In practice, ideal piece-wise steps in the intensity profile may not be realized. To run the structural change model of equation (6.1) on tiling array data, we may need to set the maximum number of segments S_{max} manually instead of relying on BIC. Figure 6.9 shows an example of the segmentation using the structural change model with S_{max} set heuristically to six. As the figure shows, sharp edges are located by the model. Without manual bounding, the BIC finds a $S_{max} = 23$, which can be seen as overfitting of the data or inadequacy of the model.

6.4   Multiscale analysis by wavelets

DNA methylation can be corroborated or induced by a single proximal methylated CpG dinucleotide via the action of maintenance and *de novo* DNA methyltransferases. Genomes thus exhibit patterns of contiguous DNA methylation. Correlational relations of DNA methylation with histone modifications help better understand the integrated regulation of chromatin accessibility in the nucleus. Because of the epigenetic modifications over extended regions of a chromosome, hybridization signals at some scale, e.g., 1 kb, may reveal better correlation than at others.

Discrete wavelet analysis [Percival00] is a powerful method of decomposing a signal y into a smooth part s_J plus a sum of details d_j at successive resolutions js,

$$y = s_J + \sum_{j=1}^{J} d_j \,, \tag{6.2}$$

where $2^J \leq N$, N being the number of data points in y. The detailed component d_j in equation (6.2) captures the changes in y on scales between 2^j and 2^{j+1} units underlying y. Missing data values in y, which is assumed to contain evenly spaced data points, can be added by linear interpolation. The top panel of Figure 6.10 illustrates a decomposition of a one-dimensional image with four pixels (9,7,3,5), the underlying units in which are one pixel. The function in d_1 in the figure is called Haar wavelet, which is dilated in d_2 and translated in d_1. The wavelet coefficients (2,1,-1) measure the degrees of "match" between the original signal and the dilated and shifted wavelets at the corresponding scale and translation. The bottom panel of Figure 6.10

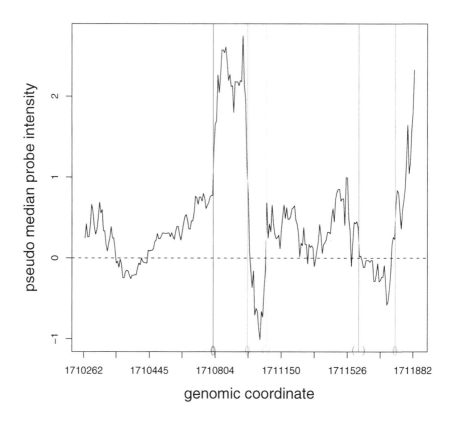

FIGURE 6.9: Segmentation of the methylation regions by the structural change model of equation (6.1) on the pseudomedian estimates of the PM−MM probe intensities of Figure 6.5. The parentheses on the x-axis indicate uncertainty in the boundary identification.

shows a similar decomposition of the intensities of hybridization of enriched fragments to a tiling array.

If we can regard the details as noise (this is the reason why we choose the wiggly Haar wavelet), the signal can be readily denoised by subtracting the details from the signal. The denoised signal s_J, therefore, smooths the fluctuations with period shorter than 2^{J+1} units. The smooth curve s_5 in the bottom panel of Figure 6.10 shows such a denoising of DNA methylation by wavelet analysis. We may perform the denoising up to different scales. Figure 6.11 shows the heatmap of s_J of the hybridization data of Figure 6.10 for J equal to 0, 1, 2, ..., 9. A suitable scale then can be chosen for subsequent analysis, such as array comparison. Note that biology also has a word on the suitable genomic scales. If statistical properties of the methylation profiles at many different scales are the same, the system is self-similar. Self-similarity may provide clues to the mechanical mechanisms behind DNA methylation.

6.5   Unsupervised segmentation by hidden Markov model

The wavelet-based multiscale analysis of tiling array data determines a scale and also smooths the data. We then are prompted with the question of whether the genomic region is "on" or "off." A simple thresholding on the smoothed intensities may determine the state of the genomic region, but the choice of a threshold value is subjective. The on or off hidden state at a region is to be inferred from the measured hybridization intensity at the same region. Suppose the probability of observing a methylation intensity when the state is on (or off) is known and can be described by a, say, Gaussian with mean and standard deviation. These probabilities are called emission probability. Suppose the state at region i depends only on that at region $i-1$. This property is called Markovian [Rabiner89]. Suppose the probability that a state at i is on (or off) given that the state at $i-1$ is on (or off) is known. These probabilities are called transition probability. With the emission and transition probabilities, the hidden state at regions i, $i = 1, 2, ..., N$, can be found from the observed methylation intensities at the same N regions by the Viterbi algorithm. That being said, we first have to estimate the parameters of the hidden Markov model including the emission, transition, and marginal probability of the state at the first region. The task is accomplished by the Baum-Welch algorithm, which successively alternates between estimation for the data likelihood given a trial set of parameters and maximization of the data likelihood by adjusting the parameter values until convergence in the parameter values. Figure 6.12 shows such a segmentation of the methylation intensities over the ENCODE ENm011 region, after wavelet smoothing, by the hidden Markov model.

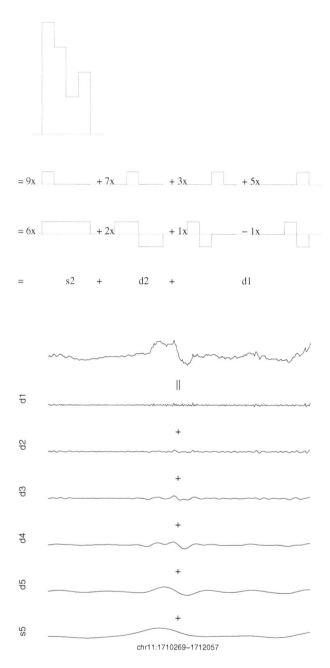

FIGURE 6.10: Wavelet decomposition of signals into smooth and detailed components. The top panel shows a one-dimensional four-pixel image and the bottom is the probe intensities in the selected region of Figure 6.9.

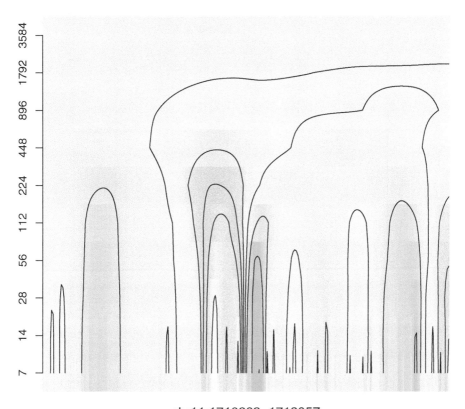

chr11:1710269–1712057

FIGURE 6.11: Peaks (light pink) and valleys (dark green) of the denoised hybridization intensities s_J of the data of Figure 6.9 up to different scales 7×2^J bp. The contours look similar across different scales, suggesting the idea of a self-similar methylation profile.

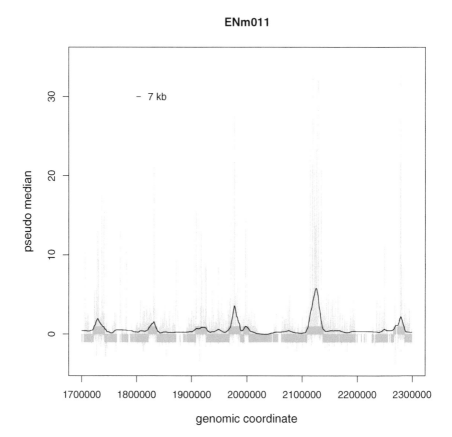

FIGURE 6.12: Categorization of the wavelet-smoothed methylation intensities s_{10} (black line) into binary states (green shade). Data, consisting of $53,953$ probe measurements, are from the tiling array of Figure 6.9.

6.6   Principal component analysis and biplot

As researchers have little idea of what to expect from a genome-wide profiling using high-density tiling arrays, it is wise to first explore the new territory with only a handful of tiling arrays. The other reason for the frugality is the array cost although it is continuously slipping. We, therefore, imagine that samples of different nature are hybridized to only a dozen tiling arrays without technical replication. The samples can be from different tissues of the same or different individuals at various disease status using different enrichment strategies, ..., and so on. The plan is to learn as much as possible from the pilot study with limited numbers of tiling arrays.

If each sample comes from a different condition in such type of exploratory studies, the number of factors (or variables) approaches that of tiling arrays, presenting a challenge for statisticians. We introduce a multivariate analysis called principal component analysis (PCA) that is suited to the current task.

An ENCODE2 tiling array has 2.3 million probe pairs, each of which can be thought of as an observation of the sample that was hybridized to it. It would be of extreme use to learn both the difference between the probes and that between the samples. Suppose there are twelve samples on twelve arrays. Twelve samples give twelve numbers for each probe, in other words creating a twelve-dimensional space for each probe. Differences in probe responses can be seen from the relative positions of the probes in the twelve-dimensional space. The difficulty arises as to visualizing the probes in a space whose dimensions are larger than three.

The twelve dimensions might not be all independent. For example, samples one, two and three may have come from the same tissue while others from various tissues. The probe readings from samples one, two and three thus are more correlated than any other trios. If we somehow combine the correlated samples and use the merged one as a replacement for the three original samples, we effectively reduce the dimensions from twelve to ten. PCA works out the combination, reducing the high dimensions to usually two for the sake of two-dimensional graphical presentation. Each of the two dimensions, called first and second principal components, incorporates probe responses from the samples, with the first principal component captures the majority of the variation in the samples and the second principal component, independent of the first component, the second majority. Points close together in the two-dimensional principal component space indicate similarity in the probe responses among the twelve samples.

On the other hand, since samples one, two and three give similar responses to the observation, their contributions to the formation of the two principal components must be similar. If we take a sample's proportional contributions to a component as a coordinate, we get a point on the two-dimensional principal component space for each sample. We draw a line linking the point and the origin. Then the three lines representing samples one, two and three must subtend small acute angles because of their similar contributions to the principal components.

It is convenient to present the two types of relations (i.e., probe–probe and sample–sample relations) on a single two-dimensional plot [Zhang07]. The result is called biplot of PCA. Figure 6.13 shows such an example. From the lines in the biplot, we see that the samples can be grouped into three clusters corresponding to three tissue types. In addition, for those points that are along a group of lines of, say, spleen tissue, we can say that the methylation status at the genomic positions of those probes represent the epigenetic (i.e., DNA methylation) markers of spleen. A PCA biplot analysis, therefore, uncovers probe–probe, sample–sample and probe–sample relations.

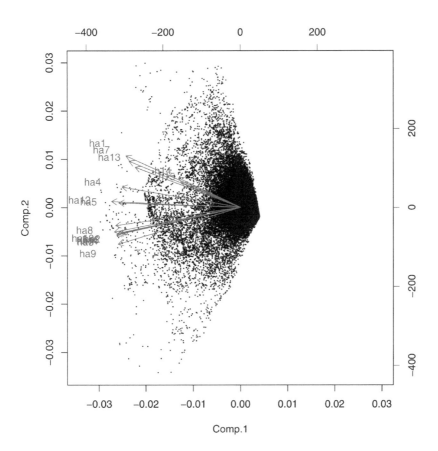

FIGURE 6.13:   Biplot of principal component analysis result. Black dots indicate probes (or observations) and red arrows indicate samples. Dots close together mean their responses are similar across the samples. Arrows close together mean the samples are correlated. Dots along an arrow, if any, can be thought of as the DNA methylation markers of the sample.

Chapter 7

Cluster Analysis

As the number of CGI (CpG island) loci in a high throughput microarray goes up, the methylation in previously unexplored and unannotated genomic regions is measured. One way to characterize the unannotated CpG islands is to associate them with the annotated CpG islands in the microarray. The function of the unannotated CGIs and the downstream genes can be implicated from that of the associated annotated CGIs. The association is accomplished by clustering, which is the topic of this chapter.

When associating loci with other loci, as in the example above, we talk about clustering on loci. On the other hand, occasions arise when we associate biological samples. For example, the methylation of colon cancer cells is profiled and we are associating arrays into clusters. Different clusters of arrays may reflect cancer at different stages of progression. In this application, we refer to clustering on samples. The clustering algorithms in this chapter are applicable to either application.

We first introduce principal component analysis, which is useful for graphical presentation of high dimensional data on two dimensions. Clustering is considered an exploratory analysis, which generates rather than proves hypotheses. In particular, different clustering algorithms yield different clustering results. If we understand the various premises and (dis)advantages of different clustering algorithms, we can better choose the algorithm that is appropriate for the application. We introduce procedures to assess the quality, significance and reproducibility of clusters.

7.1 Measure of dissimilarity

Before clustering, we need to decide a metric of dissimilarity between two objects (loci, genes or samples). The most commonly used dissimilarity measure in cluster analysis is Euclidean distance, d_{12},

$$d_{12} = \sqrt{\sum_{i=1}^{p}(x_{1i} - x_{2i})^2} \, , \qquad (7.1)$$

where x_{1i} is the methylation level (log intensity for single-color oligonucleotide arrays or log ratio for two-color DNA microarrays) of locus 1 at condition (or sample) i and there are p conditions in the experiment. When exploring microarray data with clustering, we are generally interested in the shape of the methylation profile, but not in the actual magnitude of the methylation levels. For example, loci that show similar methylation across conditions are considered closely related to each other irrespective of whether the ratios between them are 1.2 or 2.0. The problem with Euclidean distance equation (7.1) is that its values depends on the shape as well as magnitude of the profile. One remedy is to set the methylation mean to zero and standard deviation to one by the following standardization before clustering,

$$x_{1i} \leftarrow \frac{x_{1i} - \mu_1}{\sigma_1} , \qquad (7.2)$$

where μ_1 and σ_1 are the mean and standard deviation of the methylation of locus 1 across conditions.

The other common dissimilarity measure is 1 minus Pearson correlation co-efficient. Pearson coefficient measures the strength and direction of a linear relationship (i.e., correlation) between two sets of data points. Pearson coefficient is independent of the magnitude. Also it can be shown that 1 minus Pearson correlation coefficient is proportional to squared Euclidean distance on standardized datasets. After standardization, results of clustering using 1 minus Pearson correlation are equivalent to those using Euclidean distance.

7.2   Dimensionality reduction

Thousands of probe sequences reside on a typical microarray. On the other hand, up to a hundred or so samples can be hybridized in a single microarray experiment. We take as an example a time course experiment where samples after radioactive exposure are measured at five different time points by CpG island microarrays, each of which contains 13,056 spots. Figure 7.1 shows the log methylation ratios over time. In the figure, a row represents a locus and a column represents a microarray. The color is applied in such a way that the lighter (darker) the color, the larger (smaller) the log ratio. In the case of missing data, nothing is painted, leaving a blank (i.e., white) cell. Missing data points can occur when background intensities are larger than foreground intensities, giving rise to unacceptable negative intensities. Now, suppose the goal is to find methylation patterns that are characteristic to different classes of samples. Plots like Figure 7.1 give us few clues as a sample is represented by an overwhelming number (i.e., 13,056) of attributes. On the other hand, if the aim is to find characteristic profiles across different experimental conditions (or phenotypes), the large number of conditions can

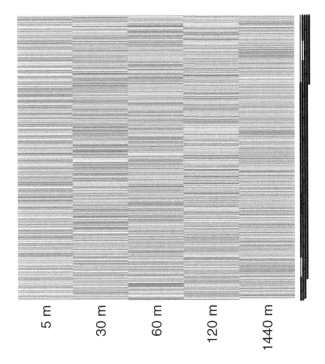

5 m 30 m 60 m 120 m 1440 m

FIGURE 7.1: Heatmap of a time-series DNA methylation data. Different rows are for the different probes on the microarray and columns are at different time points in minutes since radiation treatment.

slow down or even prohibit convergence of our estimate of the profiles to the true profiles.

Microarray data are conveniently stored in a matrix where rows record loci and columns conditions (cf. Figure 7.1). If the number of loci is n and the number of conditions is p, the dimension of the matrix is $n \times p$. For the purpose of visualizing clusters, we would like to have a way of projecting the n p-dimensional vectors (or p n-dimensional vectors) into a two or three dimensional space.

A solution to the curse of dimensionality is to find a lower dimensional space in which minimum error due to the projection is produced while maximum variability in the data is preserved. A most commonly used technique for dimension reduction is principal component analysis (PCA), which is also known in linear algebra as Karhunen-Loève transformation. Imagine the log intensities or log ratios of the ith sample are stored in a column vector $\vec{a}_i = (a_{i1}, a_{i2}, \cdots, a_{in})^T$, where n is the number of spots on the microarray slide and superscript ^{T} stands for matrix transpose. We average samples to get $\bar{\vec{a}} = (1/p) \sum_{i=1}^{p} \vec{a}_i$, where p is the number of samples (or phenotypes) in the experiment. We then subtract $\bar{\vec{a}}$ from individual samples to get $\vec{a}'_i = \vec{a}_i - \bar{\vec{a}}$

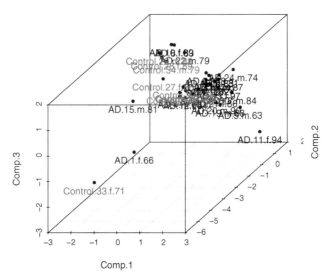

FIGURE 7.2: Data in the first three principal component space.

and form the $n \times n$ sample covariance matrix Σ by,

$$\Sigma = \tfrac{1}{n-1} \left(\vec{a}_1' \; \vec{a}_2' \cdot\cdot \; \vec{a}_p' \right) \left(\vec{a}_1' \; \vec{a}_2' \cdot\cdot \; \vec{a}_p' \right)^T$$

$$= \tfrac{1}{n-1} A A^T \, . \tag{7.3}$$

The components $\vec{u}$'s are constructed by solving for the eigenvalues and eigenvectors of the covariance matrix Σ,

$$(\Sigma - \lambda_i I)\vec{u}_i = 0, \quad i = 1, 2, \cdots, n \, . \tag{7.4}$$

Here "eigen" originates from German, meaning "of itself". The important properties of the eigen-space of Σ are that the $\vec{u}$'s form a complete set of orthogonal bases and that λ_i is the variation in direction $\vec{u}_i$. The ratio $\lambda_i / \sum_{i=1}^n \lambda_i$ thus gives us the proportion of variation in direction $\vec{u}_i$. We order the eigenvalues so that $\lambda_1 > \lambda_2 > \cdots > \lambda_n$ and call the first few corresponding eigenvectors principal components. (Note that if the dimension of AA^T is daunting, we can instead first find the eigenvectors $\vec{v}_i$'s of $A^T A$. $\vec{u}_i$ can then be obtained by $A\vec{v}_i$ with the same corresponding eigenvalue.) Figure 7.2 shows an example of dimension reduction where the methylation at one hundred twenty-four cytosines of a sample is represented by the first three principal components of the one hundred twenty-four methylation measurements.

Besides algebra, biology can help us in the problem with dimension reduction. Although methylation (or expression) levels are profiled on a genomic scale, only a fraction of the loci (genes) are expected to hypo- or hypermethy-late (under- or over-express) in the context of the experiment. The majority of the loci (genes) are intact; their methylation (expression) levels relative to the controls remain fixed and close to 1 across different conditions such as time points throughout out the experiment. Data normalization relies heavily on the same assumption. Therefore, we can calculate the variance of the methy-lation (expression) levels, i.e., log intensities for single-channel oligonucleotide chips or log ratios for dual-channel microarrays in a reference design, across conditions. We rank the loci (genes) according to their variances and pick up, say, the top 5 percent loci (genes) from the ranked list. The dimension is in this way reduced by a factor of twenty. In addition to variance, we can also include covariance in the ranking. This amounts to ranking according to the row sum of the covariance matrix Σ of equation (7.3).

7.3   Hierarchical clustering

Hierarchical clustering results in loci (or samples) ordered in a tree-like structure called dendrogram [Kaufman90]. The structure appeals to biolo-gists by providing a view of the relatedness of loci within and between clusters [Eisen98]. There are two approaches to hierarchical clustering. Agglomerative hierarchical clustering works bottom-up, while divisive hierarchical clustering works top-down. Because the algorithm of hierarchical clustering is determin-istic and sensitive to outliers, agglomerative (divisive) hierarchical clustering generates clusters that are more accurate in the bottom (top) levels. Depend-ing on the interest of the biologist, she, therefore, should choose the approach that best suits her needs.

7.3.1   Bottom-up approach

In the beginning of an agglomerative hierarchical clustering, every locus is a cluster. The algorithm then finds and merges the two clusters whose distance is shortest. The centroid of the clusters that were just merged is calculated and to be used in between-cluster distance calculation later. The agglomerating process continues until all loci are merged into a final cluster. Variants of the algorithm exist as to how to define and calculate the between-cluster distance. For example, we pair one member from one cluster with a member from the other cluster. The pair-wise distance is then calculated. The distance between the two clusters is then defined as the largest pair-wise distance. This is called complete-linkage hierarchical clustering.

Figure 7.3 shows the agglomerative hierarchical clustering on the top twenty variable loci from the time course experiment of Figure 7.1. The scale on the

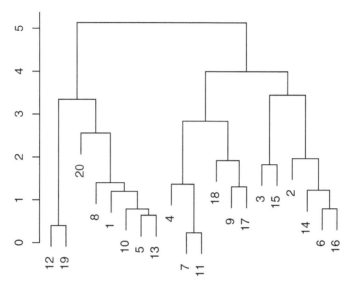

FIGURE 7.3:   Agglomerative hierarchical clustering on the top twenty variable loci of the data in Figure 7.1. The height of the shoulder measures the Euclidean distance between the clusters.

left of the tree measures the distance values. Looking at the height of the shoulder of the bifurcation, we see that locus 7 and locus 11 are closest to each other. They thus were merged in the first iteration of the algorithm. The merged loci formed a cluster that was merged with locus 4 later in the algorithm. Likewise, locus 12 and locus 19 were merged in the second iteration and loci 5 and 13 in the third iteration. The merging proceeded until it reached the top of the tree. If the methylation measurement of loci 5 and/or 13 was noisy by chance, the noise went into the centroid of their cluster. Noise accumulated along the way, deteriorating the clustering toward the top of the tree. Care has to be taken in interpreting results of hierarchical clustering.

How many clusters there are is a tough question to many clustering algorithms. In hierarchical clustering, the answer can be relegated to biologists. If, based on prior knowledge, we believe there are only two clusters, we cut the tree at the appropriate level to get two subtrees. If on the other hand, we opt for a three-cluster solution, we cut the tree at a bit lower level to get three independent subtrees. The cutting procedure is shown in Figure 7.4. (More about where to cut the tree will be provided later.) Note that clusters are nested in the dendrogram; clusters reside within a cluster. Also noted is the fact that two loci next to each other in the tree does not necessary mean they are similar. For example, locus 4 and locus 13 in Figure 7.3 are right next to each other. But they belong to two big clusters the distance between which is as large as 5.

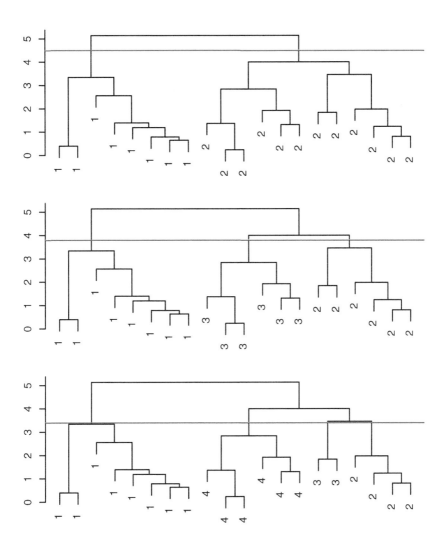

FIGURE 7.4: Cutting of the dendrogram at different levels, giving rise to different numbers of clusters. The red horizontal line at 4.5 in the top panel cuts the tree into two subtrees, the loci within each of which belong to a cluster. When we lower the cut, more clusters are generated, as indicated by the labels.

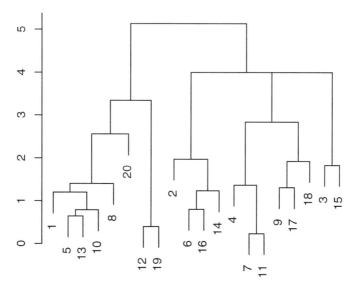

FIGURE 7.5: Result of divisive hierarchical clustering. The input data are the same as Figure 7.3

7.3.2 Top-down approach

Divisive hierarchical clustering works the opposite way. All loci are assigned to a single cluster to begin with. The locus that has the largest average pairwise distance to the other loci is picked up and split from the original cluster, forming a new cluster. Loci that are closer to this new cluster than the old cluster change their association to join the new one. This divides the cluster into two clusters. The procedure stops when every locus is itself a cluster. We show in Figure 7.5 the result of divisive hierarchical clustering on the same twenty loci as those in Figure 7.3. Figure 7.6 displays clusters resulting from different cuts on the divisive tree.

Comparing Figure 7.3 and Figure 7.5, we see that the agreement between the two approaches is not bad. In particular, memberships of some of the clusters are identical. Because the algorithm starts from the top levels of the tree, the divisive approach in principal better captures the gross structure of the tree. If the biologist has no presumption about the number of clusters, the number of steep stems may give us clues. Long stems correspond to large differences in cluster distances. So, for example, in Figure 7.5, we may be tempted to cut the tree at the height of 3, giving rise to five clusters. Figure 7.7 shows the clustering on both loci and time points with the results arranged in a heatmap where different methylation levels are represented by colors.

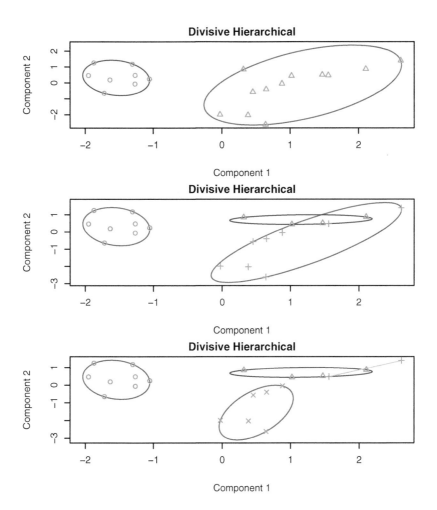

FIGURE 7.6: Results of different cutting of the Figure 7.5 tree displayed on its two-dimensional principal component space. Loci in a cluster are enclosed in a loop.

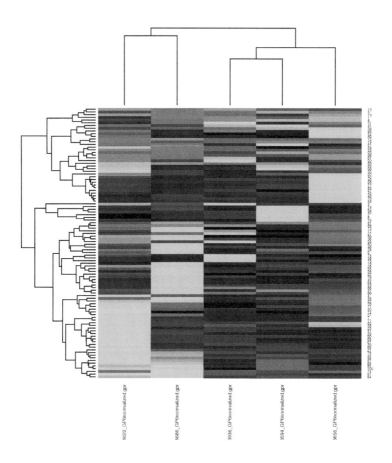

FIGURE 7.7:   Heatmap of the top one hundred variable methylation loci over the five time points. Also shown are the hierarchical clustering on both loci and conditions (i.e., time-points). Note that the order of the time points and loci are rearranged according to the clustering results. With the clustering, we do see patterns of methylation in the two-dimensional space.

7.4   K-means clustering

K-means is another most widely used clustering technique. In the method, the number of clusters, K, is an input number specified by the user. The method then tries to minimize the sum of squared distances of loci g from the centroids $\bar{C}_i(g)$ of the clusters that they belong to,

$$\arg\min_{\{C_i\}} \bar{D}(\{C_i\}) = \arg\min_{\{C_i\}} \frac{1}{n} \sum_{i=1}^{K} \sum_{g \in C_i} |g - \bar{C}_i(g)|^2 . \qquad (7.5)$$

The algorithm starts from K randomly selected cluster center locations. Every locus finds its distances to the K centers and associates itself to the nearest cluster. After the associations, the cluster center locations are updated. The processes of locus association and center updating alternate and repeat until the center locations change no more. The algorithm is not deterministic in that different initiating centers can yield different results. Therefore, the algorithm is usually run several times with random initialization. The result that gives the smallest sum of squared distances is taken as the final. The other option is to initialize K-means with the partition result from a hierarchical clustering. Figure 7.8 shows the K-means clustering on the same top twenty loci as in the previous hierarchical clustering examples.

The number of clusters has to be given to run a K-means, which poses no difficulty if we know the number in advance. In many situations, however, we do not know it. If we increase the number of clusters, the sum of squared distances, equation (7.5), decreases. When the number of clusters reaches the maximum equal to the number of loci, the sum goes down to zero as each locus sits on its center. We need to stop increasing K at some point. A principled approach is Bayesian information criterion (BIC), which amounts to maximizing the probability $P(K, \{C_i\}|\{\vec{g}\})$ of model (K and $\{C_i\}$ in our case) given DNA methylation data ($\{\vec{g}\}$),

$$\begin{aligned}
\text{BIC} &= \arg\max_{K,\{C_i\}} \log(P(K, \{C_i\}|\{\vec{g}\})) \\
&\simeq \arg\max_{K,\{C_i\}} \log(P(\{\vec{g}\}|K, \{C_i\})) - \frac{K \cdot p}{2} \cdot \log n \qquad (7.6) \\
&= -\frac{n}{2} \arg\min_{K,\{C_i\}} \log(\bar{D}(K, \{C_i\})) - \frac{K \cdot p}{2} \cdot \log n + \text{const} .
\end{aligned}$$

The higher the BIC score, the better the model. The first term on the right-hand side of equation (7.6) is, apart from a scale factor, the log of the least mean squared distances of equation (7.5). The second term is a penalty because of the minus sign. It is proportional to $K \cdot p$, which equals the number of parameters (i.e., the coordinates of cluster center locations) in the model. In other words, BIC finds a parsimonious model that best fits the data. Note

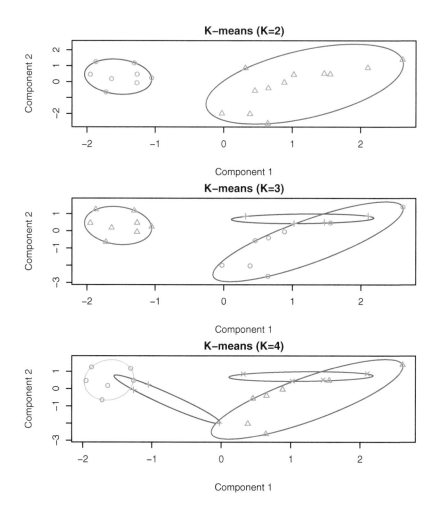

FIGURE 7.8:   Results of K-means clustering, on the same twenty top variable loci as those in Figure 7.5 and Figure 7.6, with different input K. Different
clusters are labeled with different symbols and enclosed in different loops.

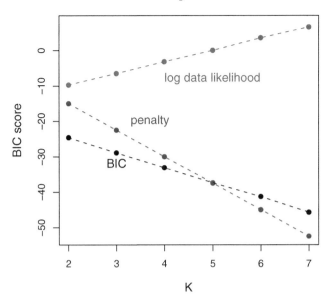

FIGURE 7.9: BIC score versus number of clusters in K-means clustering.

that the role of n and p can change depending on what we are clustering on. If we cluster on samples, in equation (7.6), n is the number of samples and p the number of loci on the microarray.

In Figure 7.9, we plot the BIC scores versus K. The gain in the data likelihood (i.e., the first term of the BIC) as K increases does not overcome the penalty. It, therefore, suggests that $K = 2$ is a better choice for the number of clusters in clustering the twenty most variable loci in the time course experiment. K-means clustering tends to find clusters of loci that spherically distribute with equal volume in the high (p- or n- depending on what we are clustering on) dimensional space. This is due to the property of the objective function that the algorithm tries to optimize: The weights to the within-cluster sums of squared distances are the same for different clusters. As we can imagine that loci in a biological pathway may be more tightly regulated than loci in the other pathway, a clustering method, such as the model-based clustering algorithm to be described below, that makes the allowance is desirable.

7.5 Model-based clustering

DNA methylation is correlated with transcriptional silencing. Loci that show similar methylation profiles across conditions may regulate the expression of genes of similar functions. Co-methylation thus can result from co-regulation in an epigenetic regulatory network. The expression level of a gene in the upper hierarchy of a biological network can be critical in initiating the expression of downstream genes. The expression level of a gene could be stringent at one stage and less so at the other stages of development. Or, the gene expression levels of one cluster can have more leeway than those of the other clusters. Methylation profiles with unequal variances between conditions thus can be envisioned. Moreover, a promoter, when collaborating with other promoters, can be involved in different genome activities.

The biological picture suggests that the methylation (or expression) profiles of different clusters of loci (genes) should be distributed with varying orientations, shapes and volumes in the high (e.g., # of time points in a time course experiment) dimensional space. Instead of weighting clusters equally, model-based clustering accommodates unequal weights between clusters in equation (7.5). The result is a spherical distribution of methylation profiles with unequal volume in the high-dimensional space. If we allow common unequal weighting of profiles within each cluster, the distribution is elliptical with equal volume, shape and orientation between clusters. The most unconstrained model is unequal weights between and within clusters. Model-based clustering achieves the parametrization of weights via the covariance matrix of a mixture of multivariate Gaussian distributions. Furthermore, the likelihood of the mixture model is taken as the joint probability of a locus (gene) belonging to a certain cluster.

The more general a model is, the more parameters it needs. The number of data points naturally places a cap on the complexity of the model that can be built. In analogy to equation (7.6) for K-means, the model-based approach to clustering employs Bayesian information criterion on the issue of model selection. Figure 7.10 shows the clustering of the twenty loci in the radiation time course experiment by the model-based method. The result returns a best model that has five spherical clusters with unequal volume.

7.6   Quality of clustering

Clustering purports to maximize intracluster cohesion and intercluster separation. However, whatever data are given, clustering algorithms return clusters. Clustering thus is more a hypothesis generating than a hypothesis testing tool in microarray data analysis. It is desirable that we have a measure of quality for the overall clustering result as well as for each returned cluster. Suppose that locus i belongs to cluster C, we describe the so-called silhouette

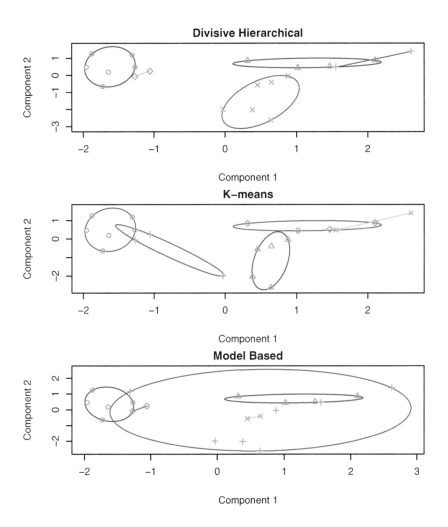

FIGURE 7.10: Different clustering results from different clustering algorithms.

width S_i that quantifies the clustering quality of locus i,

$$S_i = \frac{b_i - a_i}{\max(b_i, a_i)},\qquad(7.7)$$

where a_i is the average distance of locus i to all the other loci in cluster C, and b_i is the smallest of the distances between locus i and all the loci that do not belong to cluster C. Possible values of the silhouette width are within the range: $-1 < S_i < 1$. Note that the smaller the a_i, the tighter is locus i to the cluster. As a_i approaches zero, S_i approaches the maximum 1, indicating good clustering of locus i. The smaller the b_i, the closer is locus i to its neighboring cluster. As b_i approaches zero, S_i approaches the minimum -1, suggesting wrong association of locus i with cluster C. After calculating the silhouette width of each locus, we can form the silhouette width of a cluster by averaging the silhouette widths of the loci belonging to the same cluster. Likewise, we can find an overall silhouette width of the clustering by forming the average of all the silhouette widths. Figure 7.11 displays the silhouette widths of individual loci by divisive hierarchical clustering with different subtree cutting.

7.7 Statistically significance of clusters

Silhouette width serves as a quantitative measure for cluster quality. More often we want to assess the statistical significance of a cluster. We describe a procedure to estimate the p-value of a cluster by going through the ritual of statistical hypothesis test, which involves the following three steps: (1) define a test statistic for cluster homogeneity, (2) formulate the null hypothesis and null distribution, and (3) obtain the p-value of the cluster.

The average distance of pairs of loci in a cluster is easy to calculate and thus serves as an appropriate statistic for the test of homogeneity of loci in the cluster. The null hypothesis of the test then states that the average distance is zero. The null distribution that corresponds to the null hypothesis can be defined by setting to zero the mean methylation of every locus in the cluster while preserving the covariance structure of the DNA methylation data. We then calculate the statistic from a resample of the null distribution. After, say, a thousand resamplings, we get a distribution of a thousand average distances. The one-tailed p-value of the cluster, whose average distance is denoted by d, can be obtained by dividing by one thousand the number of times the average distances from the resampled null distributions exceeds the true average distance d.

The procedure outlined above applies to clustering on loci. When the clustering is on samples, we notice that the mean zeroing step in getting the null

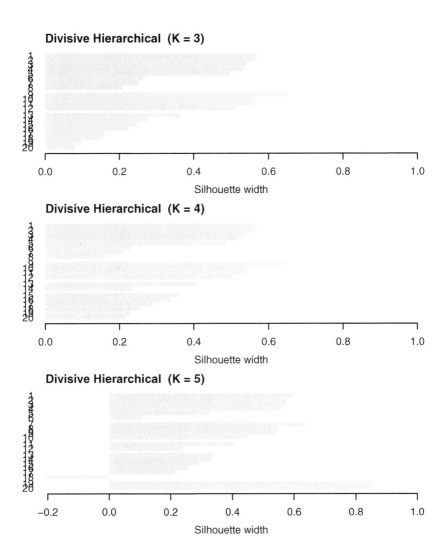

FIGURE 7.11: Quality of clusters by their silhouette widths.

distribution does not work because the distribution of DNA methylation is already centered by normalization (cf. chapter 4). Fortunately, we seldom use all the loci on the microarray in clustering. Instead, we select loci which show great variability across samples/conditions. (If we could select the loci by, say, two-sample t-test, meaning that we have already known the classes of the samples, we would not be doing the clustering.)

The above procedure estimates the p-values of individual clusters. A clustering yields multiple clusters. We can adjust the p-values for multiple testing by either Bonferroni or false discovery rate methods discussed in section 5.5.

7.8   Reproducibility of clusters

An issue that is closely related to significance is reproducibility (or stability) of clustering results. DNA methylation and gene expression is a stochastic process. Measurement of methylation or expression levels is further corrupted by noise. Loci are grouped together in a cluster by a clustering algorithm using mean methylation levels. Because of the uncertainty in the mean, how often will the same loci be found grouped together?

We utilize simulation to address the issue. In linear model analysis of DNA methylation or gene expression data, mean methylation or expression levels (i.e., log intensities or log ratios) are represented by fitted coefficients. Uncertainties (i.e., errors) in the coefficients are returned as the mean sum of residuals. We then can generate an artificial dataset by adding, to each coefficient, a resample of the error. We filter loci by picking out the, say, the top one hundred variable loci. We then apply the clustering algorithm to the simulated methylation of the selected loci, and obtain a clustering result. We repeat the resampling, filtering and clustering, say, one thousand times. We then count the numbers of times pairs of loci are in the same cluster among the one thousand clustering results. Finally, we report clustering result at a, say, 95 percent reproducibility with the loci that co-appear over 95 percent of the time in the clusters.

7.9   Repeated measurements

We have so far assumed that mean values of replicated data points are used for distance calculations in clustering algorithms. The use of mean values is straightforward and applicable to most clustering algorithms. It, however, does not take full advantage of information from repeated measurements, which provide not only mean values, but also standard deviations of the means.

An approach to improve the reproducibility of clustering using repeated measurements is to down-weight noisy data during distance calculations,

$$d_{12} = \sqrt{\sum_{i=1}^{p} \frac{(\bar{x}_{1i} - \bar{x}_{2i})^2}{\sigma_{1i}^2 + \sigma_{2i}^2}} , \qquad (7.8)$$

where $\bar{x}_{1i}$ is the mean methylation level (log intensity or log ratio) of locus 1 at condition (e.g., phenotype) i: $\bar{x}_{1i} = (1/r)\sum_{j=1}^{r} x_{1ij}$, and σ_{1i} is the standard deviation of x_{1ij} over the $j = 1, 2, \cdots, r$ repeated measurements of samples. If samples are to be clustered and r spots are replicated on the microarray, d_{12} is the distance between samples 1 and 2, the summation is over the p selected loci, and the means and standard deviations are over the replicated spots on the microarray. If the replicate measurements on sample i is noisy, i.e., $\sigma_{1i}^2 + \sigma_{2i}^2$ is large, the contribution of sample i to d_{12}, among the total of p samples, is diminished according to equation (7.8).

Chapter 8

Statistical Classification

One of the major applications of DNA methylation microarrays is to identify epigenetic markers for disease diagnosis. Other applications include classifying diseased samples into distinct subtypes. Classification algorithms are widely used for pattern recognition, which is one of the main subjects in machine learning. A familiar example is to build a spam filter that classifies incoming e-mails into spam and nonspam.

High-density microarrays measure the methylation at CGI (CpG island) locations across entire genome. Not all of the locations are informative in classification as not all the words in an e-mail are useful in discriminating the e-mail. The first step in building a classifier thus is to select the informative loci. The step is critical as we select discriminatory keywords in successfully filtering e-mails.

The building of a classifier relies on a training dataset in which the disease status of every sample is known. Parameters of the classifier are then tuned in order to minimize the classification error. It is usually found that simple classifiers perform well in comparison to sophisticated ones. We introduce two simple classifiers and also illustrate the performance of a classifier by the technique of receiver-operating-characteristic curves.

8.1   Feature selection

Classification is not new. It is used in tasks such as document categorization and face recognition. DNA sample classification with microarray data is, however, of particular challenge because of the problems with small sample size and noise that are inherent in microarray experiments. A large-scale microarray experiment can involve up to hundreds of sample-array hybridizations. A typical microarray, however, features tens of thousands of spots on a single slide. In DNA methylation, it is know that hypermethylation of the CpG islands in or around the promoters of tumor suppressor genes is associated with cancers. Silencing of different tumor suppressor genes in different tissues may lead to different cancers. In gene expression, it is estimated that only about 40 percent of the genes are expressed in a cell at a particular instance of time. Many of them are so-called housekeeping genes that are constantly expressed in order to maintain the basic operation of the cell. We, therefore, can say that most of the data points (log intensities or log ratios) are uninformative and uninteresting because of their irrelevance to phenotype discrimination. The uninteresting loci show no differential methylation/expression across samples. If the uninteresting loci are not removed from the analysis, not only do they aggravate computational load, but they may contribute to misclassification due to measurement error.

Biological knowledge helps in selecting loci for classifier building. For example, if we know *a priori* the sequences that code for the proteins (e.g., tumor suppressors) in the biological machinery (i.e., expression silencing during carcinogenesis) that characterizes the biological (cancerous) samples in the microarray experiment, we should include the promoters in the learning set of the classification. Unfortunately, oftentimes, we do not have the knowledge, which explains why we launch genomic-wide measurements using microarrays for clues. Chapter 5 and chapter 6 serve the purpose and describe methods to identify interesting loci. Specifically, t- and F-tests and their non-parametric counterparts, Wilcoxon tests, are introduced to rank differentially methylated loci across conditions in terms of their statistical significance. In the same spirit, we can try to rank loci according to the signal-to-noise ratio, SNR,

$$\text{SNR} = \frac{|\hat{\mu}_1 - \hat{\mu}_2|}{\hat{\sigma}_1 + \hat{\sigma}_2} , \tag{8.1}$$

where $\hat{\mu}_i$ and $\hat{\sigma}_i$ are, respectively, the sample mean and sample standard deviation of the methylation in class i.

After ranking by a test-statistic, we need to decide how many loci to be included in the training set for classifier building. It is believed that the number of samples per class should ideally be five to ten times the number of loci per class in developing a classifier. Since the number of samples per class can be up to around one hundred, we are forced to select about ten or less loci per class for classification. In Figure 8.1, we show the selected features on a PCA (principal component analysis) biplot. In the example, the methylation at one hundred twenty-four CpG-dinucleotide sites in the promoters of eleven genes were measured from thirty-four brains, out of which twenty-four were Alzheimer's disease patients. The five roughly equally spaced blue arrows in the plot indicate that they work independently and synergistically to represent the thirty-four samples. The five sites were among the top loci by site-wise two-sample t-tests. The other one hundred nineteen CpG sites are not informative and need not be included in classification.

Loci (genes) selected according to the rank may end up coming from the same group of loci (genes) that are co-methylated (co-regulated) by a DNA methyltransferase (transcription factor). Because we want to include as many decorrelated loci as possible for the purpose of class discrimination, we can first cluster loci and then select the few top ranking loci from each cluster in the hope that a diverse army of loci are selected to represent a class. Chapter

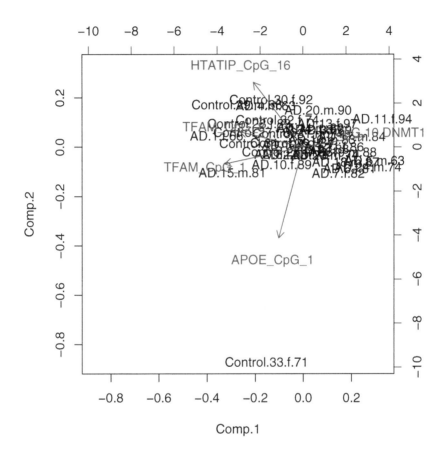

FIGURE 8.1: Biplot of PCA on one hundred twenty-four cytosine sites. Black labels indicate the thirty-four brains whose methylation were measured at the one hundred twenty-four sites. The blue arrows indicate the contributions of the five sites in the first two principal components of the methylation measurement.

7 covers several popular clustering algorithms.

8.2 Discriminant function

Classification is a statistical decision theory problem and Bayes statistics is convenient in tackling the issue [Berger85]. Suppose $p(k, x)$ is the joint probability of class k and data x and that $p(x)$ is the probability of data $x = (x_1, x_2, \cdots, x_n)$ where x_i is the methylation (or expression) at locus i and n is the number of informative loci selected for the classification study. Then the conditional probability $p(k|x) = p(k, x)/p(x)$ is the probability of class k given the data x. By the equality $p(k, x) = p(k|x) p(x) = p(x|k) p(k)$, Bayes theorem states that, $p(k|x) = p(x|k) p(k)/p(x)$. Because $p(x)$ is fixed given the observation of x, it serves as a normalization constant to the probability. We can write the Bayes theorem as,

$$p(k|x) = \frac{p(x|k) p(k)}{\sum_{l=1}^{K} p(x|l) p(l)} \propto p(x|k) p(k) , \qquad (8.2)$$

where there are K different classes in the samples. $p(k|x)$ is also called class posterior probability because we calculate it after data are taken. Similarly, $p(k)$ is called class prior because we somehow know it before observing the data. $p(x|k)$, class-conditional probability, is the density of x in class k. With the Bayes rule that minimizes misclassification, classification of data x amounts to finding the class k that is maximal among the (log) posterior probabilities,

$$
\begin{aligned}
C(x) &= \arg \max_{k} p(k|x) \\
&= \arg \max_{k} p(x|k) p(k) \\
&= \arg \max_{k} \log(p(x|k) p(k)) .
\end{aligned}
\qquad (8.3)
$$

With the statistical skeleton, different assumptions about the distribution of data points in the class, i.e., different parametrizations of the class-conditional density $p(x|k)$, lead to variants of classifiers. For example, a general form of the class-conditional density is the multivariate Gaussian, thanks to the central limit theorem,

$$p(x|k) = \frac{1}{(2\pi)^{n/2} |\Sigma_k|^{1/2}} \exp\left(-\frac{1}{2} (x - \mu_k)^T \Sigma_k^{-1} (x - \mu_k) \right) , \qquad (8.4)$$

where μ_k is the mean vector and Σ_k the covariance matrix of the methylation (expression) of the loci (genes) in class k. Now we collect the methylation data from samples with known class labels (e.g., various cancers) to form the so-called training set (uninformative loci are thrown out according to the previous section). We then estimate the class mean μ_k and class covariance Σ_k from the data in the training set, x', by,

$$\mu_k \leftarrow \hat{\mu}_k = \frac{1}{N_k} \sum_{i \in k} x'_i$$

$$\Sigma_k \leftarrow \hat{\Sigma}_k = \frac{1}{N_k - 1} \sum_{i \in k} (x'_i - \hat{\mu}_k)(x'_i - \hat{\mu}_k)^T,$$
(8.5)

where the summation is over the samples in class k and N_k is the number of samples in that class. The class prior $p(k)$ can be estimated from the proportion of cases in class k among all the N samples,

$$p(k) = \frac{N_k}{N} .$$
(8.6)

With the estimation for the prior (8.6) and the distributional assumption (8.4) with parameter estimates (8.5) from the training samples, the class membership of new methylation (expression) data x is predicted by the discriminant function (8.3),

$$\hat{C}(x) = \arg \max_k \left[-\frac{1}{2} \log |\hat{\Sigma}_k| - \frac{1}{2}(x - \hat{\mu}_k)^T \hat{\Sigma}_k^{-1}(x - \hat{\mu}_k) + \log N_k \right], \quad (8.7)$$

where terms independent of k have been dropped. The hat on $C(x)$ emphasizes the fact that the prediction is based on quantities estimated from the samples in the training set. The major term, i.e., the second term, on the right-hand side of equation (8.7) is the (squared) Mahalanobis distance between the two vectors x and $\hat{\mu}_k$. An intuitive interpretation of the discriminant function (8.7) then states that the predicted class of x is the class whose estimated class mean vector $\hat{\mu}_k$ is, under an appropriate distance metric, closest to x.

8.2.1 Linear discriminant analysis

We make the simplifying assumption that the covariance matrices are the same for samples of different classes: $\Sigma_k = \Sigma$. We also estimate Σ from the covariance matrix of all samples,

$$\Sigma_k \leftarrow \Sigma \leftarrow \hat{\Sigma} = \frac{1}{N - K} \sum_{k=1}^{K} \sum_{i \in k} (x'_i - \hat{\mu}_k)(x'_i - \hat{\mu}_k)^T.$$
(8.8)

Plugging it into equation (8.7) and again ignoring terms that have nothing to do with k, we find that the discriminant function becomes

$$\hat{C}(x) = \arg \max_k \left[x^T \hat{\Sigma}^{-1} \hat{\mu}_k - \frac{1}{2} \hat{\mu}_k^T \hat{\Sigma}^{-1} \hat{\mu}_k + \log N_k \right]$$

$$= \arg \max_k \delta_k(x) .$$
(8.9)

Note that the general discriminant function (8.7) is quadratic in x but with the simplification of a common covariance matrix of equation (8.8), it becomes linear. As a consequence, the boundary between any two classes k and l is a plane in the n-dimensional space and can be obtained by solving for the xs that satisfy the equation $\delta_k(x) = \delta_l(x)$.

Classification by equation (8.9) is called linear discriminant analysis. Compared with the elliptical class boundaries of the quadratic discriminant analysis equation (8.7), linear boundaries of the linear discriminant analysis are less flexible. However, because less parameters need be estimated, classification by linear discriminant analysis appears in practice more robust.

8.2.2   Diagonal linear discriminant analysis

In the linear discriminant analysis (8.9), we may ignore correlation, assuming that the covariance matrix is diagonal:

$$\Sigma = \mathrm{diag}(\sigma_1^2, \sigma_2^2, \cdots, \sigma_n^2) \leftarrow \mathrm{diag}(\hat{\sigma}_1^2, \hat{\sigma}_2^2, \cdots, \hat{\sigma}_n^2)\,, \qquad (8.10)$$

where we have again used sample quantities to approximate population quantities. The classification rule then becomes

$$\hat{C}(x) = \arg\min_k \left[\frac{1}{2} \sum_{i=1}^n \left(\frac{x_i - \hat{\mu}_{k_i}}{\hat{\sigma}_i} \right)^2 - \log N_k \right]. \qquad (8.11)$$

Simplifying assumptions greatly reduce the number of probability density parameters to be estimated. Although strong assumptions can introduce bias to the true distribution of DNA methylation, error in classification $\hat{C}(x)$ can be less because of better estimates for the parameters. Simpler classifiers thus are found to perform well in practice. Performance, however, also depends critically on what loci are selected for classification.

8.3   K-nearest neighbor

In contrast to the parametric approach that leads to a discriminant function, k-nearest neighbor (kNN) are a nonparametric classifier. In this case, again, a distance metric is used and kNN uses either Euclidean distance or one minus correlation coefficient. k is a positive integer in kNN. Suppose k equals 3 and there are two classes in the training samples. Then the classification rules of the 3NN are: (1) among the training set, find the three samples whose distances to the unknown sample are shortest; and (2) the class of the unknown sample is then decided by the class of the majority of the three samples. So, for example, if two (or three) out the three samples belong to class one, then the class of the unknown sample is class one. Since class prediction is by majority vote, k is better chosen from odd integers for dichotomous classification. Ties still happen if the number of classes is greater than two. In these cases, they are broken at random.

In the classification by kNN, the class-conditional probability density $p(x|k)$ is approximated by

$$p(x|k) = \frac{k_V}{N_k V} \,, \tag{8.12}$$

where V is a volume element around x in the n-dimensional space, k_V is the number of samples of class k in V and N_k is the total number of samples of class k in the training set. The prior (unconditional) density $p(x)$ of data x is estimated by

$$p(x) = \frac{\sum_{k=1}^{K} k_V}{NV} \,, \tag{8.13}$$

where the numerator is the total number of samples in volume V, K is the number of classes, and N is the total number of samples in the training set. Putting equation (8.12), equation (8.13) and equation (8.6) into Bayes theorem gives immediately the kNN rules

$$p(k|x) = \frac{p(x|k)\,p(k)}{p(x)} = \frac{k_V}{\sum_{k=1}^{K} k_V} \,. \tag{8.14}$$

A small integer k in kNN translates to a small volume element V. In other words, the smaller the k, the local-er information of samples around x is used to predict x's class. The class boundaries in kNN are nonlinear. The larger the k, the smoother the boundary. A larger k takes into account more samples around x, and, in addition, could provide richer probabilistic information among the competing predictions (cf. equation (8.14)). On the other hand, too large a k is detrimental because it destroys the locality by considering too many distant, irrelevant samples. Optimal values of k can be determined by cross validation.

8.4   Performance assessment

A classifier is a function that maps a data vector into a class label. To build a classifier, we use a training set containing data with known labels. However, because of noise in, and random sampling of, the training data, the predicted class is a random variable. An essential quantity of a classifier thus is its accuracy, or 1 − error rate. Ideally, after training, we use yet the other independent dataset, whose class labels are also known, to test the performance of the classifier. The dataset is called test set. In microarray data

TABLE 8.1: Linear Discriminant Analysis, # of Loci = 20

		Predicted	
		Schizophrenia	Bipolar Disorder
Actual	Schizophrenia	7	5
	Bipolar Disorder	8	4

classification, numbers of samples are already small. If we further divide the limited samples into training set and test set, we suffer more from the small sample size problem. In practice, therefore, we use the training set for both training and cross validation. As it was pointed out that feature selection was part of the classification, loci should be selected using the training set. If they are selected with the help of the whole data, misclassification rates will be downward estimated.

It is found that complicated classifiers, which perform well with plenty of samples, may not perform as well as simple classifiers (such as diagonal linear discriminant analysis and k-nearest neighbor) when numbers of samples are small. This is because a small variance in prediction, compared with bias in prediction, often leads to an overall small classification error.

8.4.1   Leave-one-out cross validation

A way to use training set for both training and validation is the leave-one-out cross validation (LOOCV). In the scheme, a sample is first taken out of the training set. The class of the taken sample is then tested by the rest of the samples in the training set. The leave-one-out process is done one sample by one sample until all the samples in the training set are tested for their class labels.

As an example, we consider an experiment profiling methylation of twenty-four diseased cases versus a pool of healthy controls. After normalization, we rank the loci by t-statistic. Twenty top loci are selected. The cases are actually from two subtypes: twelve schizophrenia and twelve bipolar disorder. We train a classifier to see how well the subtype is classified by the selected loci. Table 8.1 shows the performance of a linear discriminant analysis by leave-one-out cross validation. The classification error rate is the proportion of the sum of the off-diagonal counts over the total counts in the table: (5 + 8)/24. We see that the rate is as bad as a random guess. Indeed, we test the null hypothesis of no association between the actual and predicted classes against the alternative hypothesis of a positive association by Fisher's exact test for count data. The p-value of the Fisher's test on Table 8.1 is found to be 0.8 and we fail to reject the null hypothesis. We suspect that the poor performance might be due to the noise in the selected loci. We reduce the

TABLE 8.2: Linear Discriminant Analysis, # of Loci = 5

		Predicted	
		Schizophrenia	Bipolar Disorder
Actual	Schizophrenia	11	1
	Bipolar Disorder	2	10

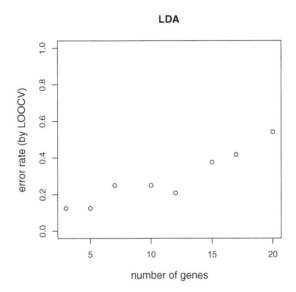

FIGURE 8.2: Error rate of a linear discriminant analysis classifier as a function of the number of selected loci.

number of loci and repeat the assessment by cross validation. Figure 8.2 shows the classification error rate as a function of number of loci. We see that the error does drop. When the number of selected loci is five, the classification result is shown in Table 8.2. The p-value of Fisher's test on Table 8.2 is 0.00032, suggesting a significant association of prediction with true class.

We can use the same technique to find the optimal k for k-nearest neighbor. As an illustration, we use the same training set (twenty-four diseased samples) as in the linear discriminant analysis above. Borrowing from the previous result, we use the top five loci for kNN. We repeat the leave-one-out cross validation for different ks. The result is shown in Figure 8.3, which indicates that k around 7 gives a better performance. Table 8.3 lists the result of the 7-nearest neighbor.

TABLE 8.3: Result of 7-Nearest Neighbor, # of Loci = 5

		Predicted	
		Schizophrenia	Bipolar Disorder
Actual	Schizophrenia	11	1
	Bipolar Disorder	1	11

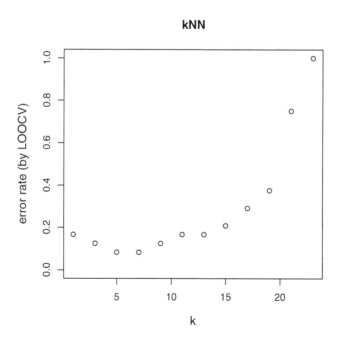

FIGURE 8.3: Error rate of a kNN classifier as a function of k given the top five loci from Figure 8.2.

TABLE 8.4: Confusion Matrix

		Test Result	
		Negative	Positive
Actual Status	Normal	a	b
	Diseased	c	d

8.4.2 Receiver operating characteristic analysis

When assessing the performance of a diagnostic classifier by, say, leave-one-out cross validation, we obtain the so-called confusion matrix of Table 8.4. The false positive rate is calculated by the ratio $b/(a + b)$ from the matrix. Since specificity is the chance of identifying nondisease, i.e., $a/(a + b)$, the false positive rate is equal to $(1 - \text{specificity})$. The true positive rate is by the ratio $d/(c + d)$. Since sensitivity is the chance of detecting disease, the true positive rate is equal to the sensitivity. An ideal classifier has a false positive rate of 0 and true positive rate of 1. If we plot the false positive rate on the x-coordinate and the true positive rate on the y-coordinate, we obtain a receiver operating characteristic (ROC) point. For each classifier, we obtain a point. The better classifier is then the one whose point is closer to (0,1) on the plot.

To better understand the utility of ROC, we summarize the task of a kNN with leave-one-out cross validation classification in an example: (1) the training set consists of thirty-four brains whose classes (i.e., Alzheimer's (AD) or controls) are known; (2) for brain i, its distances to the other thirty-three brains are calculated; (3) the k brains, which are closest to brain i are identified; (4) the class of brain i is then determined by the majority of the k brains' classes. Now, each brain's methylation was measured at one hundred twenty-four cytosine sites and the questions are: (1) which methylation sites are to be included in the classification and (2) k = ? True positive claims on the twenty-four AD brains increase at the cost of increasing false positive claims on the ten control brains. The ideal scenario is to have high true positive rate and low false positive rate. In Figure 8.4, we show the determination of both the informative sites and k using the ROC curve.

In many situations, the costs of misdiagnosing positives and negatives are not symmetric. For example, if a person is diagnosed to be ill when she is, in fact, healthy, the cost of this false positive is further tests for the illness. On the other hand, if one is diagnosed to be healthy when she is, in fact, ill, the cost of the false negative can be fatal. When comparing and optimizing classifiers, we often need to take into account the relative cost of misclassification. ROC curves help achieve the goal. Let α and β denote, respectively, the cost of a false alarm and missing a positive. The total cost of both types

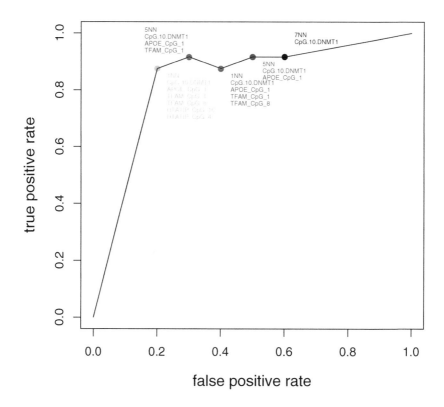

FIGURE 8.4: ROC curve of kNN classification with leave-one-out cross validation. The performance of the classifier varies as the selected sites and/or value of k change. If accuracy is defined as $(tp+tn)/(p+n)$, then the red classifier of 5NN with three marker sites and the green classifier of 1NN with six marker sites give, respectively, an accuracy of $(22+7)/(24+10) = 0.85$ and $(21+8)/(24+10) = 0.85$. That is, they perform equally well in prediction.

of misclassification is then

$$\text{cost} = \frac{1}{a+b+c+d} \cdot (\alpha \cdot b + \beta \cdot c)$$

$$= (1 - \frac{c+d}{a+b+c+d}) \cdot \alpha \cdot \frac{b}{a+b} + \frac{c+d}{a+b+c+d} \cdot \beta \cdot (1 - \frac{d}{c+d}) \qquad (8.15)$$

$$= (1 - p) \cdot \alpha \cdot x + p \cdot \beta \cdot (1 - y) \, ,$$

where x is the false positive rate, y the true positive rate, and $p = (c+d)/(a+ b+c+d)$ the proportion of positives in the training set. An optimal classifier is selected to be the one whose cost is least. Equation (8.15) indicates that equo-cost points form parallel, straight lines and that to lower the cost we decrease x and increase y. As an example, Figure 8.5 shows the ROC curve of a kNN classifier with different k. Suppose there are equal numbers of diseased cases and normal controls in the training samples, i.e., $p = 0.5$. If the cost of a false positive is the same as that of a false negative, i.e., $\alpha = \beta$, the slope of the equo-cost lines is 1. Examples are the green lines in Figure 8.5. To choose the number of nearest neighbors for the kNN classifier in this case, we see that k = 5 or 7 are better than k = 3 and 9. On the other hand, if $p = 0.5$ and $\beta/\alpha = 5$, i.e., the cost of a false negative is five times that of a false positive, the slope of the equo-cost lines is 1/5. Examples of such asymmetric cost cases are the red lines in Figure 8.5. We see that the optimal choice for the k in kNN is k = 9 in this case. Note that the ruggedness of the ROC curve (in blue) in Figure 8.5 is due to small sample size. If the number of samples in the training set can be increased, the rates can be estimated at a higher resolution, resulting in a smoother ROC curve and a better determination of the value of k.

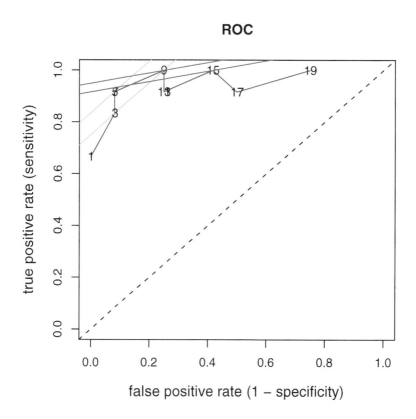

FIGURE 8.5:   ROC curve and the equo-cost lines of a kNN classifier.

Chapter 9

Interdependency Network of DNA Methylation

The number of human CpG islands was estimated to be within a hundred thousand. One approach to the genomic scale data analysis has been to perform locus-wise statistical test for methylation differences between cases and controls. After correcting for multiple testing, we identify candidate loci for verification by independent techniques. Although extension to the standard tests has been developed to enhance the power of the tests, the issue of interdependence in the between-locus DNA methylation has not been systematically addressed.

Silencing of the genes in a pathway can happen during development, leading to specific tissues. Other examples of gene silencing via DNA methylation include genomic imprinting and X chromosome inactivation. Knowledge of the inter-locus DNA co-methylation is important. Construction of the map of DNA co-methylation is a step ahead, accounting for the interdependence unaccounted for by conventional approaches.

We introduce "net" correlation as a measure of pairwise co-methylation. The pairwise relation is direct in the sense that we calculate the Pearson correlation between two methylation loci after taking into account the effects of the methylation at other loci. Statistically significant relations are then conveniently represented by edges linking nodes, forming a graph (or network). The topology of such graphs is most likely shaped by evolution. We introduce metrics to characterize the DNA co-methylation networks. Studies of the mechanisms underlying the emergent structure of the co-methylation graphs help better understand epigenetics. In particular, comparisons of the healthy with diseased graphs elucidate what has gone wrong in development and/or mitosis.

As the number of CGI loci in a network can be huge, we partition the network into subnetworks in such a way that edges within the subnetworks are denser than those across subnetworks. The identified subnetworks are called modules. The loci in a module may correspond to a suite of regulatory elements that are involved in controlling a pathway. We thus go on to test for the hypotheses that particular chromosomes and roles (e.g., gene ontology, pathways and protein domains) are enriched in the modules. A graph approach helps better understand epigenomics, providing insights into innovative pharmaceutical, environmental and behavioral interventions to disorders.

9.1   Graphs and networks

Methods in chapter 5 for the identification of differentially methylated loci run statistical tests on arrays one locus at a time. We get p-values from individual tests and adjust the p-values for multiple testing. We then rank the loci for significance according to the adjusted p-values. The procedure assumes independence of the methylation status at a locus from another. The assumption may be over-simplistic as we know that cells undergo tissue-specific DNA methylation during early organismal development. Aberrant *de novo* DNA methylation over regions of genome has also been observed throughout stages of tumorigenesis. The methylation states at some loci, therefore, are interrelated. As a step beyond the locus-wise test for differential methylation of chapter 5 and chapter 6, we introduce graph theoretical method to take into account the interdependency of methylation among loci.

A graph (or network) is composed of nodes and edges. A node represents a variable and in the context of DNA methylation microarray experiment, is the measured methylation status at a locus. The status can be the mean log intensity on a probe from one-color microarrays or the mean log intensity ratio at a probe from two-color microarrays in a common reference design. If the correlation coefficient of methylation between two nodes across samples, such as measurements at different time points in a time-series experiment, is found significantly large, we draw an edge connecting the two nodes. Since we establish an edge based on correlation, which is invariant under a ↔ b, i.e. `cor(a,b) = cor(b,a)`, the edges are undirected. If the establishment of an edge involves regression models, which explicate the response of a node to the other, we can assign directionality to the edge so that the graph becomes directed. We focus the discussion on undirected graphs in this chapter and on directed graphs in the next chapter.

9.2   Partial correlation

When there are only two variables in a system, correlation by Pearson product-moment coefficient introduced in chapter 1 is useful for the relationship between the two. When there are more than two variables in the system, correlation between two variables may be mediated by a third one. Since indirect relations can be obtained once we acknowledge all the direct relations, indirect relations are redundant. The primary goal, therefore, is to uncover the direct relations, which can be measured through partial correlation.

Suppose the methylation g_i at locus i is measured by a microarray investigating a total of N loci. The relation of DNA methylation between locus 1 and the rest excluding locus 2 is modeled by

$$g_1 = \alpha_3 g_3 + \alpha_4 g_4 + \cdots + \alpha_N g_N + \epsilon_1 , \qquad (9.1)$$

where α_i are regression coefficients and ϵ_1 is the residual. We write down the relation of DNA methylation between locus 2 and the rest excluding locus 1 in a similar fashion,

$$g_2 = \beta_3 g_3 + \beta_4 g_4 + \cdots + \beta_N g_N + \epsilon_2 , \qquad (9.2)$$

where β_i are another set of regression coefficients. After regressing equation (9.1) and equation (9.2), which amounts to minimizing ϵ_i by finding the best α_i and β_i, the partial correlation between loci 1 and 2 is calculated by the Pearson product moment coefficient between the two minimized residuals: $V_{12} = \mathrm{cor}(\epsilon_1, \epsilon_2) = V_{21}$. The procedure is applied to any pair i, j, yielding $N(N-1)/2$ partial correlations between any pair of loci whose methylation are measured by the microarray.

Note that it can be shown [Lezon06] that the probability of a measured methylation profile $p(\mathbf{g}) = p(g_1, g_2, \cdots, g_N)$ gives rise to a maximum entropy S defined as

$$S = -\sum_{\mathbf{g}} p(\mathbf{g}) \log(p(\mathbf{g})) \qquad (9.3)$$

when

$$p(\mathbf{g}) \sim e^{-1/2 \sum_{ij} g_i V_{ij} g_j} \qquad (9.4)$$

with V_{ij} being the partial correlation between g_i and g_j and proper centering of the data: $< g_i >= 0$. Partial correlation, therefore, quantifies the interaction between two DNA methylation loci. The profile $p(\mathbf{g})$ in equation (9.4) is also the most probable given the samples.

9.3   Dependence networks from DNA methylation microarrays

The linear regression model like equation (9.1) has $N - 2$ unknown parameters. To estimate them, we would need on the order of N different samples. Since N, the number of probes (i.e., loci) in a microarray, can be tens of thousands, it is impossible to have that many samples and microarrays. We would in practice have to preselect a subset of loci from the microarray.

If the methylation at a locus is constant across individuals, so is the residual according to equation (9.1). Thus, the locus is not likely to correlate with other loci in terms of methylation in the samples. Therefore, we would want to discard the loci showing low variations in methylation across individuals. In other words, we calculate and rank the standard deviation of the methylation intensities (or ratios) across the samples for every probe on the microarray and select, say, the top one hundred variable loci for partial correlation calculation. We can calculate the mean methylation of the rest of the loci on the microarray and designate the mean to an artificial locus called "rest." Every microarray is then represented by the $101 = (\text{top } 100 + \text{rest})$ measurements. The N is effectively diminished to 101, a manageable size.

If we have two groups of samples, one being affected individuals and the other unaffected controls, we calculate and rank the standard deviations of each group. We then select top one hundred loci from each group and form the union of the two top one hundred lists. If forty loci appear in both lists, the union will then consist of one hundred sixty unique loci. We then use the one hundred sixty loci in the construction of the dependence network of DNA methylation for each group of samples. Likewise, we may follow the "mean field" technique above and use one hundred sixty-one loci.

Similar to the p-value for a Pearson product-moment correlation coefficient (cf. equation (1.19) in section 1.2.7), a p-value associated with the test for null hypothesis of zero partial correlation can be obtained [Schafer05]. Among the $N = 100$ selected loci, we obtain the $100 \times 99/2$ partial correlations and associated p-values. We can adjust the p-values for multiple testing by the method of false discovery rate (FDR) of section 5.5.2 and establish an edge between any two loci whose partial correlation is significantly different from zero. If we set the significance level at 0.01 and, say, 800 partial correlations have their FDR-adjusted p-values smaller than 0.01, then 8 among the 800 edges are expected to be false positives.

Alternatively, we permute the one hundred preselected probes in each array. The model of (9.1), (9.2), $\cdots$, etc. is then applied to the permuted data to obtain $100 \times 99/2$ partial correlations and the associated p-values. The smallest (unadjusted) p-value can then be used as the cutoff for the p-values from unpermuted data. The rationale is that since the relationships among the probe measurements are destroyed and become random by permutation, any genuine partial correlation should be more significant than the best that would be obtained by chance. Suppose we choose the p-value cutoff to be 10^{-7}. From each permutation, we get the number of false positive edges whose p-values are lower than 10^{-7}. After many permutations, we have a sense of the average false positive edges in the network under the 10^{-7} cutoff.

In Figure 9.1 is shown the dependence network of DNA methylation from eighteen control sperm on fifty-four CpG island microarrays in a common reference design. An edge between two nodes means a direct association of methylation between the two loci observed among the sperm samples. That is, an increase in methylation at one locus is associated with an increase (or decrease) in methylation at the other locus among normal sperm.

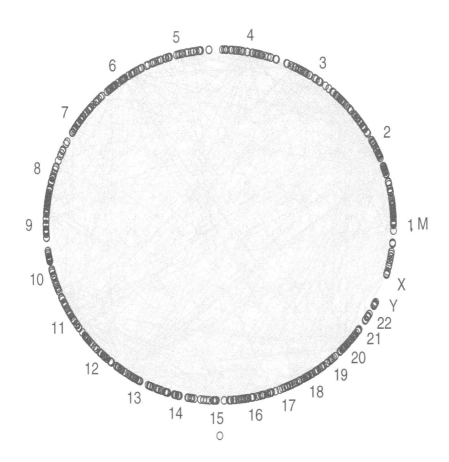

FIGURE 9.1: Interdependence network of DNA methylation using CpG island microarrays. 1079 CGI loci whose methylation levels are variable among the eighteen individuals are selected in network reconstruction. An edge is drawn between two loci if the partial correlation is significantly (raw p-value $< 10^{-7}$) different from zero. The genomic locations of the loci are mapped to the human genome, which is arranged around a circle. A single locus right underneath the circle is the "rest" representing the mean methylation of the rest of the selected loci. (See text for detail.)

9.4 Network analysis

Biological networks [Barabasi04], such as metabolic networks, protein–protein interaction networks, gene transcription regulatory networks and neural networks, have been studied and shown to exhibit properties that are not found in random networks, in which edges are randomly assigned to pairs of nodes. For example, we may calculate the shortest distance between a substrate and a product in a metabolic network by counting the number of reactions the substrate takes to produce the product along a metabolic pathway. We obtain the distances for any two nodes in the network and get the mean of all the pair-wise distances. To compare, we create a random network by detaching the edges and reassigning them to nodes that are randomly picked and paired. We again calculate the mean distance of the random network. It was found that many networks in nature including biological networks have shorter mean distances than random networks. A shorter mean distance in the case of metabolic (neural) networks may reflect an evolutionary legacy since metabolites (excitations) are produced (received) with greater efficiency.

In addition to mean distance, many metrics that are defined on the network as a whole or on the individual nodes are informative. We can rank the nodes according to the metric and select the top ranking nodes for further investigation. We can also pick up the nodes whose metrics from the methylation network of affected samples are very different from those of control samples. The network approach is to complement the traditional analysis of differential methylation, which ignores the interrelations among loci.

Before moving on, we introduce the adjacency matrix that conveniently represents a network. For a network with N nodes, we create an N by N matrix A. The (i, j) element of A is one, $A_{ij} = 1$, if nodes i and j are connected by an edge and zero, $A_{ij} = 0$, otherwise. Note that for an undirected network, A is symmetric: $A_{ij} = A_{ji}$ and that, by the definition of partial correlation equation (9.1), the diagonal elements of the matrix are zero: $A_{ii} = 0$. The number of edges in a network is the number of ones in the matrix divided by two. The average number of edges per node, therefore, is

$$< k >= \frac{1}{2N} \sum_{i=1}^{N} \sum_{j=1}^{N} A_{ij} \, , \tag{9.5}$$

which quantifies how densely the nodes in the network are interconnected.

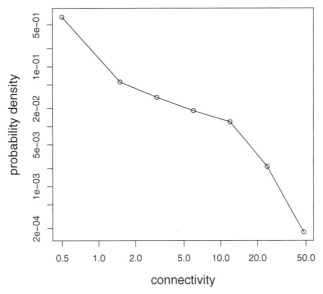

FIGURE 9.2: Frequency distribution of the connectivities (# of edges per node) of the network in Figure 9.1. Note that both axes are in logarithmic scale. The average number of edges per node $<k>$ of the network is 1.5.

9.4.1 Distribution of connectivities

The number of edges k attaching to a node can be readily obtained by

$$k_i = \sum_{j=1}^{N} A_{ij} \ , \tag{9.6}$$

for all the nodes $i = 1, 2, \cdots, N$ in the network. The frequency distribution $p(k)$ of connectivity k gives the chance of picking up a node whose connectivity is equal to k.

For a random network, the connectivities are Poisson distributed, peaking at average number of edges per node $<k>$. A distribution of connectivities that deviates from a Poisson indicates that the network is nonrandom. Figure 9.2 shows the connectivity distribution of the network of Figure 9.1, the non-Poisson of which suggests a nonrandom component in the formation of the network.

9.4.2 Active epigenetically regulated loci

The fat tail on the right of the connectivity distribution of Figure 9.2 indicates enrichment of nodes with large k. High-k nodes are those that connect to many other nodes; they are hubs in the network. Hubs in a protein-protein interaction network are thought to be essential proteins since they can bind with many other proteins to form complexes that assume functional roles in many pathways. Mutations in the DNA sequences coding for the hubs thus can be lethal. Likewise, hubs in the dependence network of DNA methylation may coincide with hot regulatory regions (such as DNaseI hypersensitive sites, histone modification sites or transcription factor binding sites) including CGI loci and promoters, the methylation of which correlate with that of many other regulatory regions.

Given two methylation networks, one from controls and the other from cases, we compare lists of the top connectivity nodes in the networks and identify the nodes that are different in terms of the ranks. For example, a locus may be present in one of the top lists, but absent in the other. The loci that are identified may be actively involved in pathways in one group of individuals. The same pathways, however, are relatively dormant in the other group of individuals. The method of network analysis, therefore, has the potential of identifying shifts in the functional paradigms (e.g., pathways) between groups of individuals.

9.4.3   Correlation of connectivities

A next step in network analysis is to look at the average connectivity of a node's connected nodes, which can be calculated by

$$k_{nn\,i} = \frac{1}{k_i} \sum_{j=1}^{N} A_{ij} k_j \, , \qquad\qquad (9.7)$$

and is called assortativity $k_{nn\,i}$ of node i. We find the assortativities of all the nodes in the network and plot $k_{nn\,i}$ versus k_i. If the assortativities increases (decreases) with connectivities in the plot, it indicates that hubs tend (not) to connect with other hubs in the network. The network is called assortative (disassortative).

Human social networks are assortative in the sense that people of the same background, such as language and race, live together, just like birds of a feather flock together. Technological networks, such as the World Wide Web, are disassortative because of the competition for Web visitors among the giant Web servers, such as Yahoo and Google. In Figure 9.3, we plot the average assortativities K_{nn} of the nodes having the same connectivities over the range of the connectivities, $k = 1, 2, \cdots, k_{max}$,

$$K_{nn}(k) = \frac{1}{N_k} \sum_{i=i}^{N} k_{nn\,i} \delta_{k_i, k} \, , \qquad\qquad (9.8)$$

where $N_k = \sum_{i=1}^{N} \delta_{k_i,k}$ is the number of nodes whose number of edges is k and $\delta_{k_i,k} = 1$ if $k_i = k$ and $\delta_{k_i,k} = 0$ otherwise. A decrease in K_{nn} with k in Figure 9.3 indicates that the methylation at hot CGI loci are more correlated with cold CGI loci than with other hot CGI loci.

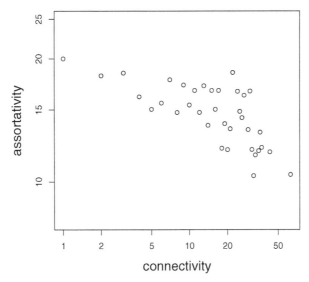

FIGURE 9.3:   Frequency distribution of the assortativity (neighbor's mean # of edges) of the network in Figure 9.1. Note that both axes are in logarithmic scale.

If we imagine the chromosomes in the nucleus as relatively immobile noodles with highly diffusive DNA methyltransferases and histone modifying enzymes moving around, the fact that the dependence network of DNA methylation is neither assortative nor disassortative may indicate that the distribution of active CGI loci is uniform in the three-dimensional nuclear space and that active CGI loci compete for no DNA methyltransferases because there are plenty of them. Changes in the assortativity of the network can implicate changes in the chromosomal organization or availability of DNA methyltransferases in the nucleus.

9.4.4   Modularity

Metabolic networks can be partitioned into subnetworks or modules so that reactions most often take place between the substrates within modules. Proteins in modules of a protein-protein interaction network may bind together to form various complexes. Modularity confers evolutionary advantage in that new exogenous functions are inserted into an evolving genome without endogenously reinventing the functions. Furthermore, complex functions can emerge from assemblies of simple functions.

To identify modules in a DNA methylation network, we try various partitions of the network so that the densities of edges between the nodes within sought modules are larger than expected by chance. There may appear different partitioning of the network that fulfills the requirement so it helps to define network modularity Q [Newman06] that serves to compare different partitions,

$$Q = \frac{\sum_{i,j \in \text{modules}} (A_{ij} - P_{ij})}{\sum_i k_i} , \qquad (9.9)$$

where P_{ij} is the probability that nodes i,j are connected by chance. The denominator is simply a conventional normalization. Since node i has k_i edges and node j has k_j edges, a natural choice for P_{ij} is $P_{ij} \propto k_i k_j$. The constraint $\sum_{ij} P_{ij} = \sum_i k_i$ finally gives $P_{ij} = k_i k_j / \sum_i k_i$. Note that the modules in the summation in equation (9.9) are not known beforehand. The task is to find a segregation of the nodes into modules that maximizes the modularity Q in equation (9.9).

An efficient algorithm has been developed for the detection of modules in networks based on leading eigenvalue in maximizing equation (9.9) [Newman06]. The algorithm starts with the network as a single module and calculates its modularity. It then divides the network into two subnetworks, improving the modularity Q as a result. Each of the two subnetworks is subject to division. Division of the divided subnetworks goes on until the modularity Q improves no more. The indivisible (sub)subnetworks are the modules. An advantage of the algorithm is that the hierarchical structure of subsubnetworks within subnetworks is unraveled on the way toward maximizing Q. In Figure 9.4, we show the hierarchy of the identified modules in the network of Figure 9.1 using the eigenvalue based algorithm for module detection. Note that improvement in Q becomes insignificant toward the leaves of the dendrogram. A hierarchical display of Figure 9.4 showing the relations of the subnetworks is useful in case we are merging modules into megamodules. Note also that we found large modules containing lots of CGI loci as well as many small modules containing only a couple of loci. Figure 9.5 results from the redrawing of Figure 9.1 with the modular information of Figure 9.4.

9.4.4.1 Positional enrichment analysis

CGI loci in a module are more likely co-methylated than those between modules. We can hypothesize that the CGI loci in a module cluster and co-localize within a chromosome. Before testing the hypothesis, we plot the positional distribution of the selected loci along the genome in Figure 9.6. As seen in the figure, the selected loci show no enrichment in particular chromosomes. The dependence network of DNA methylation is built from the preselected loci based on the significance of partial correlations under a preset false discovery rate.

We count the number E_i of network loci that locate in chromosome i and plot the distribution of the counts versus chromosomes in Figure 9.7. Similarly, for a module, we count the number O_i of loci in the module that are located in chromosome i to obtain the chromosomal distribution of module loci. The null hypothesis of the positional enrichment test states that the

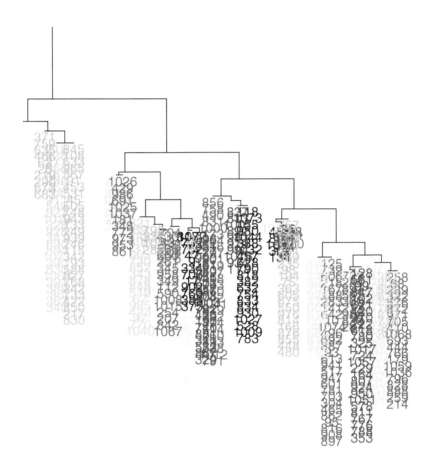

FIGURE 9.4:   The hierarchical structure of the DNA methylation network of Figure 9.1 by the eigenvalue-based algorithm to maximize the modularity of equation (9.9). The improvement in the modularity in each division is indicated by the length of the vertical bar. Division of a subnetwork stops when there is no improvement in modularity. The indivisible leaves in the hierarchical tree are modules. The number of modules in the network is forty. The median of the module sizes is ten loci. Locus IDs are labeled by numbers.

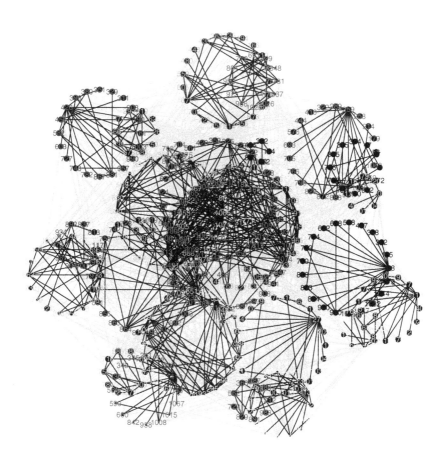

FIGURE 9.5: DNA methylation network of Figure 9.1 redrawn with the modular information in Figure 9.4. Nodes in a module are in the same color. Within-module edges are in black and between-module edges are in gray. The coloring and labeling correspond to those of Figure 9.4.

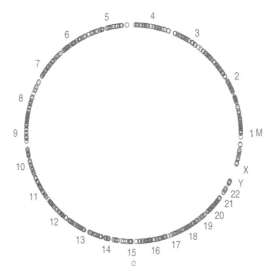

FIGURE 9.6: Distribution of the loci in a module over the genome. Genomic coordinate goes counterclockwise. Loci in red are the selected top 1079 variable loci from the methylation data. Loci in blue are in a module.

chromosomal distribution of the CGI loci in a module is no different from that of the pre-selected CPI loci.

The hypothesis can be tested by chi-square test,

$$\chi^2 = \sum_{i=1}^{23} \frac{(O_i - E_i')^2}{E_i'} \ , \tag{9.10}$$

where $E_i' = (\sum_i O_i)(E_i / \sum_i E_i)$ is what is expected from the null hypothesis. The p-value of the test can be obtained from chi-square distribution with twenty-two degrees of freedom. We run the test on each of the large modules. Any chi-square p-value below a preset significance level indicates that the loci in the module co-localize in some chromosome(s).

9.4.4.2 Functional enrichment analysis

Modularity of biological networks sounds conceptually sound. The job, therefore, is to seek evidence of association between biological functions and network modules. To this end, we first identify the genes that are downstream of the CGI loci. That is, a CGI locus on the microarray is tied to a gene. The functions of the gene can be classified according to, for example, the Gene Ontology (GO) terms the gene is mapped to. If we find an overrepresentation of a GO term for the genes in a module, we get support for the idea that the loci in a module are involved in the regulation of a biological function.

Suppose the dependence network of DNA methylation has one thousand

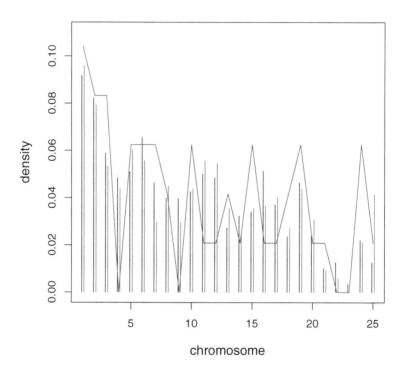

FIGURE 9.7:   Distribution of the chromosomal location of the loci. Black is the distribution of all the annotated loci in the microarray. Yellow is the distribution of the unique, nonrepetitive sequence loci in the microarray. Red is the distribution of the top 1079 variable methylation loci among the eighteen independent sperm samples. Blue is the loci in a module. P-value of the chi-square test is 0.83.

CGI loci, fifty of which are mapped to a GO term, and that one of the largest modules in the network contains sixty loci. If the loci in the module are formed from the one thousand CGI loci by chance, we would expect to find about three loci in the module to have the GO term: $60 \times 50/1000 = 3$. If instead we find thirty loci in the module having the GO term, we gain confidence that the module is not formed by chance.

The probability of drawing m apples in n picks, without replacement, from a basket of N objects containing M apples and $N - M$ nonapples is described by the hypergeometric distribution f,

$$f(m; N, M, n) = \frac{\binom{M}{m}\binom{N-M}{n-m}}{\binom{N}{n}} . \tag{9.11}$$

In the above example, $N = 1000, M = 50, n = 60$ and the one-tailed p-value of the GO term enrichment test is calculated by p-value $= \sum_{i=m}^{n} f(i; N, M, n)$. We can move on to test the other GO term, likely changing M, the number of network loci mapping to the new GO term, as a result. We then find m, the number of module loci mapping to the GO term, and calculate the p-value of m using equation (9.11). If the p-value is smaller than a predefined significance level, which is usually 0.05 divided by the number of GO terms, it is an indication that the GO term is enriched in the module. After finishing all the GO terms in the module, we move on to test GO term enrichment in another large module.

GO terms, annotating the roles of genes in organisms, are organized in a tree-like graph with root, grandparents, parents, $\cdots$, children. The annotations of children are more specific than those of parents. For example, an apple (child) is a fruit (parent). A gene mapped to a child GO term must also be mapped to the parent GO term. The GO terms thus are correlated. As a consequence, if apple is found to be enriched in the above test, we would also find that the term fruit is also enriched. Since we are more interested in specific terms, we can start the enrichment test from the children terms. Once a child GO term is found significantly enriched in the module, the loci mapped to the child GO term are removed from the list of m loci when we move to test the parent GO terms. In this way, the less specific parent GO terms are less likely to be called significant [Alexa06].

Association of module loci to KEGG pathways and PFAM protein domains can be tested in the same fashion.

Chapter 10

Time Series Experiment

De novo methylation mostly occurs in embryonic development, responsible for genomic imprinting, X-inactivation and transposon suppression in mammals. Aberrant *de novo* methylation of the CpG islands in the promoters of growth regulatory genes occurs frequently in human cancers. Maize root tissues undergo genome-wide demethylation when the seedlings are exposed to cold stress. Homeostatic imbalance of the components, such as folate, methionine and S-adenosylmethionine, in the biosynthetic pathways for DNA and histone methylation may precede disorder. A time-ordered microarray measurement of the evolving pattern of DNA methylation, monitoring the fate of 5′ cytosines, helps elucidate the regulatory program of the epigenetic machinery.

Analysis of the data by clustering and correlation helps infer the function of an unknown promoter/gene under the assumption that promoters/genes of the same biological function display similar methylation/expression profiles over time. To tap further into the data is the inference of regulatory networks that depict which promoter/gene, at what time, regulates which other promoters/genes.

We view the relations among the players in a regulatory network as a graphical model consisting of nodes and arrows (i.e., directed edges). A node, corresponding to a player, is associated with a quantity whose value is measured by microarrays or other high throughput assays. An arrow pointing from node A to node B represents that A "regulates" B in the manifestation that the observed methylation or expression of A precedes that of B. Note that for a better interpretation of the reconstructed regulatory network, quantities of different nature, such as promoter methylation, histone modifications, gene expression and protein/metabolite concentrations of the biological system, have to be measured simultaneously in an experiment.

Regulatory networks reconstructed from microarray data aim to help elucidate the mechanism of cellular processes in the molecular level. However, even for $N = 20$ selected nodes, the number of possible different network structures (combinations of arrows) is astronomical: $2^{20 \times 20} \sim 10^{120}$. Furthermore, researchers face a challenge posed by the noisy and low-replicate/frequency characteristics of current microarray data. We would prefer an approach that is parsimonious in that the number of regulatory relations in the reconstructed gene network decreases as the noise level of the expression data increases. The parsimony is to lower the false positive rate.

We describe a methodology for regulatory network reconstruction from time-series microarray experiments that aims to address the above issues. First of all, a nonlinear model is described that relates the hybridization intensity (DNA methylation or gene expression or both) of a regulated locus to a power-law function of the hybridization intensities of regulating loci. The dynamic model, in difference equation form, can be extended to accommodate delayed methylation or transcription that could arise due to transport and diffusion of gene products across cellular compartments. An objective score function is then defined that consists of a likelihood function and a penalty term. The former awards high scores to good network structures that better fit the microarray data while the latter penalizes complex structures that tend to overfit the data. The penalized likelihood score materializes the concept of parsimonious network reconstruction.

The rest is about computational implementation. Among the global search algorithms for best networks, we introduce genetic algorithms (GAs) in which a population of chromosome-like solutions to the optimization problem recombine with each other and mutate on ways toward best solution to the problem. Advantages of employing GAs are threefold. First, an observation of the objective function shows that the likelihood term approaches an asymptotic value set by noise as the network structure gets more and more complicated, i.e., more arrows. We can then design evolution in the GA such that networks grow from the simplest structure containing only one arrow. As more arrows emerge in the network, the asymptotic value is reached; further growing of the network becomes unfavorable due to the penalty term. The GA in such design effectively explores the simple structure regime, mitigating computation burden. The strategy is in accordance with the spirit that more (less) regulatory relations are reconstructed with less (more) noisy datasets.

Secondly, as the problem is usually underdetermined, many different structures may serve equally well as solutions. Network structures that survive evolution before we quit the GA are all strong candidates for the best solution to the reconstruction. Thirdly and pedagogically, in analogy to microarray measurement where sequence fragments compete to hybridize to specific probes, GAs are performing soft-microarraying where combinations of arrows compete to maximize the score.

10.1   Regulatory networks from microarray data

Signal transduction pathways are familiar examples of the cascades of reactions in the cell following external excitations. The production of the pathway end-products slows down to an eventual stop by some kind of feedback mechanisms that desensitize the receptor's response to the initiating signals. The cascade and feedback form a loop of regulation. As pathways intertwine, as exemplified by the between-module connectivities in the last chapter, the relations among the regulators (e.g., DNA methyltransferases, histone modifying enzymes, transcription factors/cofactors) and regulatees (CpG islands, core histone tails, transcription factor binding sites) form a complex regulatory network. Note that the roles of regulators and regulatees can exchange depending on the perspective. An experiment that profiles DNA methylation, histone modifications and mRNA/RNA simultaneously at different time points after stimuli has the potential of unraveling the regulatory network underlying the response. Time series experiments are also persuited by stem cell researchers for a better understanding of cellular differentiation.

We focus on time-series experiments of microarrays, in particular DNA methylation, which is the topic of the book, but other possibilities, such as an integration of the measurements, should not be disregarded. A straightforward approach to the network reconstruction is the modeling of methylation and/or expression at a locus by a linear combination of the methylation and/or expression of all the loci in the network. A total of N by N "weight" parameters will have to be determined from the N by T data points, where N is the number of loci in the network and T the number of time points in the time-series microarray experiment. Values of T range from 5 to 20, limiting the size of networks that can be reconstructed in this approach. In cases where $T < N$, solutions are underdetermined. Fortunately, an experiment is usually designed for a specific physiological process that involves no more than ∼100 genes. And after a filtering process for conspicuously methylated and/or expressed loci, only tens of loci are left.

In microarray experiments, noise arises from intrinsic stochasticity of biochemical reactions and from extrinsic errors in measurement technology, even after proper data normalization (cf. chapter 4). Uncertainty can effectively be reduced by repeated measurements. However, the numbers of technical replicates in current microarray experiments can still be as low as 1. We may curb the high false positives due to the large noise in an *ad hoc* manner with thresholds to the weight parameters. Alternatively, we will introduce an approach that casts the network reconstruction task into an optimization problem in which a score function consisting of a penalty term against complex networks is defined. We need a prescription of the model before applying the optimization.

10.2 Dynamic model of regulation

The combinatorial effect of regulator loci on regulatee locus i is modeled by the following power-law formalism,

$$x_{i,t+1} = x_{i,t} + \alpha_i \prod_{j \in S_i'} x_{j,t}^{w_{ji}} \prod_{k \in S_i''} x_{k,t-1}^{w_{ki}} - \beta_i x_{i,t} , \qquad (10.1)$$

where $x_{i,t}$, for two-color microarrays, is the hybridization intensity ratio of locus i at time t to that at time zero. α_i and β_i are positive real parameters, quantifying the methylation and demethylation rate of locus i. The products are over the sets of loci that "regulate" locus i: S_i' contains the loci whose methylation at time t affects $x_{i,t+1}$ and S_i'' contains the loci whose methylation at time $t-1$ affects $x_{i,t+1}$. The exponents w_{ji} and w_{ki} are nonzero real. If $w_{ji} > 0$, locus j "activates" locus i since $\partial x_{i,t+1}/\partial x_{j,t} > 0$. Otherwise locus j "represses" locus i. Mathematically, it was shown that many general classes of functions can be locally approximated by products of power-law functions. Empirically, it has also been demonstrated that the essential nonlinear features of synergism and saturation in many biological and complex systems can be captured by the power-law formalism [Savageau87, Savageau98]. We therefore adopt the form in equation (10.1) for a compact and tractable representation of the dynamic effect of regulators on regulatees.

In order not to clutter the formula, equation (10.1) contains only two different time delays, relative to $t+1$, on the right-hand side: t, $t-1$. It, however, is known that we can extend to include more delays if necessary. For example, we may allow four delays, t, $t-1$, $t-2$, and $t-3$, in the product on the right-hand side of equation (10.1).

10.3 A penalized likelihood score for parsimonious model

In the model of equation (10.1), xs are measured by microarrays at various time points. Now suppose we have a trial network for locus i. That is, we assume a set of S_i', S_i'', and the associated parameters α_i, β_i, w_{ji}, w_{ki} for locus i in equation (10.1). By substituting the measured fluorescent intensities at time t and $t-1$ to the right-hand side of equation (10.1), we can then predict, the intensity of locus i at the next time point, $t+1$. If our presumed network and parameter values are right, we can expect, due to noise, that the measured intensities $x_{i,t}$ deviate little from the predicted $x_{i,t}'$ and that the deviations are normally distributed: $p_{i,t}(x_{i,t}' - x_{i,t}) = N(< x_{i,t}' - x_{i,t} >, \sigma_i)$. We have assumed that the magnitude of error is independent of time, i.e.,

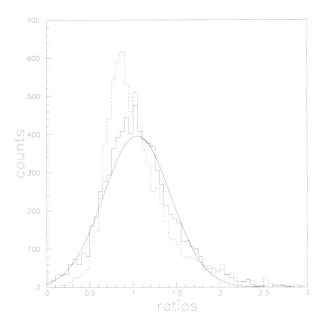

FIGURE 10.1: Distribution of expression ratios. Histograms are plotted that show the distributions of yeast gene expression ratios under heat shock. The solid line histogram shows the 8140 ratios in the time-zero dataset. The smooth curve is a Gaussian fit to it. The mean of the Gaussian is 1.04 and the standard deviation is 0.39. The dashed line histogram shows the 8140 ratios 5 minutes after the temperature upshift. Comparing the two histograms, we see that some gene expression are repressed under heat shock stress.

$\sigma_{i,t} = \sigma_i$. We do this for every time point to get a prediction of the time-ordered hybridization intensity profile of locus i. We further assume σ is the same for every locus: $\sigma_i = \sigma$, $i = 1, 2, 3, \cdots, N$. We then estimate the standard deviation of the measurement errors, σ, by the standard deviation of the distribution of intensity ratios of the probes on the microarray. The ratios are to the intensity at the first time point (i.e., time zero). An example of σ estimation from a time-series microarray experiment is shown in Figure 10.1. The probability P_i of the time-series data for locus i, can be obtained by the product, $P_i = \prod_{t=2}^{T} p_{i,t}$, where T measurements were made at time points $1, 2, \cdots, T$.

Given the data, to differentiate possible candidate networks, we can define a score function for the regulatory network of locus i as the logarithm of P_i

plus a penalty term [Schwartz78],

$$
\begin{aligned}
\text{score}_i(S_i', S_i''; \alpha_i, \beta_i, w_{ji}, w_{ki}) = \\
-\frac{\sum_{t=2}^{T}(x_{i,t}' - x_{i,t})^2}{2\sigma^2} - \frac{T-1}{2}\log(2\pi\sigma^2) \qquad (10.2) \\
-\frac{d_i}{2}\log(N \cdot (T-1)) \, ,
\end{aligned}
$$

where the second line is log likelihood $\log(P_i)$ and the third line is a penalty proportional to the number of parameters, d_i, in the network for locus i. The total score of the overall network consisting of all the N preselected loci is obtained by summing up the above score for individual loci. This decomposition has the effect of reducing the space of possible networks from $2^{N \times N}$ to $N \times 2^N$.

The score function depends on the trial network structure S_i' and S_i'' that carry parameters, α_i, β_i, w_{ji}, w_{ki}. As the structure gets complicated (more arrows), the sum of squared prediction errors (the numerator of the first term in the second line of equation (10.2)) gets small and $\log(P_i)$ approaches the asymptotic value determined by σ (second term in the second line of equation (10.2)). Further growth of arrows increments d_i and thus decreases the score. The effect is a parsimonious reconstruction of regulatory networks.

A second observation of equation (10.2) is that for noisy microarray data characterized by a large σ, $\log(P_i)$ approaches the asymptotic value quickly when new arrows are attached to the network structure. The effect is a coarser (finer) reconstruction of networks with noisier (cleaner) data.

Note that the above algorithm aims to construct regulatory relations that hold throughout the time of the experiment. The assumption is valid as experiments are designed to unravel the programmed methylation and/or expression during a process. However, as pathways are cross linked, other regulatory programs may step in in the middle of the experiment, changing the network structure. We have to be provident about the time horizon of the experiment and preselection of the N loci.

10.4 Optimization by genetic algorithms

As the number of possible networks is astronomical, exhaustive searches in the network structure space, followed by estimation of the best parameters embedded in the trial structure, consume a great deal of computation. Furthermore, computing demand is expected to scale up with N as a polynomial in N. If we place an upper limit on the number, k, of arrows leaving/entering a node, we can greatly reduce the search space and thus computation time. It, however, is speculated that a biological network can contain "super" loci (or hubs) that link to many other loci; the network structure exhibits so-called scale-free property.

To maximize the score function of equation (10.2), we propose to find the optimal structure by genetic algorithms (GA) [Holland62] and, given a trial structure S_i' and S_i'' in the GA, we maximize $\log(P_i)$ by adjusting α_i, β_i, w_{ji}, and w_{ki} parameters by the method of downhill simplex [Nelder65], which we found efficient in minimizing the sum of squared prediction errors of the power-law form in equation (10.2) [Wang04].

A genetic algorithm is a biology-inspired computational technique to find exact or quasi-exact solutions to optimization/search problems. GAs are inspired by evolutionary idea of natural selection, consisting of iterations of inheritance, mutation, recombination and selection. First of all, an initial population of possible network structures (called chromosomes) to the problem is created. The total score (cf. equation (10.2)) or fitness of each chromosome is calculated. We then rank the chromosomes according to their fitnesses. Fitter chromosomes are then selected for random mating, producing daughter chromosomes. The daughter chromosomes are also allowed to mutate. The population then enters into a new generation where scoring, selection, mating and mutation of the chromosomes take place. The procedure of survival of the fitter repeats from generation to generation until no further improvement in the average score of the population of chromosomes. We then quit the iteration, outputting the chromosome (i.e. network) with highest score as the solution to the optimization problem equation (10.2). The advantage of GA is that it exploits random search within an otherwise indefinite space for best solutions. The production of new chromosomes by crossing over two selected chromosomes during the mating operation is believed to be a way out of local minimum traps, which plague most other search algorithms. A mutation operation in the GA can also be defined to randomly generate or eliminate an arrow.

GAs are general and thus applicable in principle to any optimization problems. However, efficient GAs rely on how to encode trial solutions into "chromosomes" that are to recombine and mutate [Wang03]. For example, we can represent a network by a vector of nodes each of which specifies its incoming and outgoing arrows. Two vectors can then cross over to give rise to two daughter vectors. A node in the vector can also mutate by repointing one of its outgoing arrow to a different destination node. GAs thus are suitable for searches in the space of network structure. On the other hand, it proves inefficient to optimize by GAs the real valued α, β and w parameters in the network structure, the reason being that it is not convenient to encode real numbers into chromosome or vector-like objects. Instead, we found downhill simplex algorithm to be efficient for the optimization of α, β and w values.

Taking advantage of the properties of the score function equation (10.2), we explore the promising low-k structural regime by growing a network from the simplest structure containing only one arrow in the GA. Moreover, for a trial network, we calculate and rank the sum of squared prediction errors of each locus. The structure mutation in the GA is then operated on the locus whose sum of squared prediction errors is the largest. Namely, we play "divide (network into subnetworks for individual loci) and conquer (the worst locus)" generation by generation in the GA. Since more arrows are needed to account for a larger sum of squared prediction errors, the effect is that our reconstruction does not prevent super loci, which are regulated by many loci.

We try a population of, say, two thousand trial networks in the GA, which starts the evolution from the simplest networks, each of which contains only one randomly generated arrow (i.e., regulator-regulatee pair). To test if the GA population size is large enough, we can run the program a couple of times to see if results are reproducible. We quit the program when the average score of the population stops increasing. A network is made up of arrows. We count the numbers of the arrows appearing in the population of networks in the last GA generation. The arrows with highest frequencies represent the most likely regulatory relations among the loci.

Chapter 11

Online Annotations

DNA sequences make up the language of living cells. Understanding the cryptic words and phrases in the grammar of life processes is the quest of molecular biology. As the volume and complexity of molecular data explode in the genomic era, computerized information processing tools, such as biomedical databases and analysis software, become an absolute requirement for improved healthcare. We describe computerized systems that store and enable users to retrieve and analyze knowledge about molecular biology, genetics and biochemistry.

Researchers at different stages of biomedical investigation appreciate different levels of biological information. For example, sequence maps and expressed sequence tags (ESTs) are essential for microarray probe and polymerase chain reaction (PCR) primer design. Annotated genes' transcription start sites serve as the reference frames for epigenetic profiling. Literature abstracts abbreviate previous findings. Disease genes and mechanisms are relevant to the discussion of new findings. A gateway to the various databases is useful for meta-analysis.

Many biological experiments have been, and only, on simpler model organisms because of technical and/or ethical issues. A picture of life that is independent of species is important in that the knowledge can hopefully be easily translated to humans. Proteins, gene products, are the fundamental working and/or structural entities of living cells. Knowledge of their structures and functions is directly related to health. Metabolism involving enzymes and small molecules in pathways is well conserved across species through evolution and closely related to pharmaceutics.

11.1 Gene centric resources

11.1.1 GenBank: A nucleotide sequence database

GenBank of National Center for Biotechnology Information (NCBI), in collaboration with the European Molecular Biology Laboratory (EMBL) of the European Bioinformatics Institute (EBI) and the DNA Data Bank of Japan (DDBJ), has been receiving sequence depositions from sequencing laboratories and centers worldwide since 1993. (See: www.ncbi.nlm.nih.gov) The accepted nucleotide sequences (over 50 bp in length) are fragments of genomic DNA or mRNA that can cover single or multiple genes from over 240,000 distinct organisms. A successful sequence submission is assigned a unique and stable accession number. Annotation of the sequence is attributed to the submitting group and includes the biological information defined by GenBank Feature Table Definition. Upon receipt of the submission, GenBank processes the sequences to make sure that sequences and translations are matched, name and lineage of the organism are corrected, vectors are not contaminated, and PubMed/MEDLINE identifiers are added. To compare a query sequence, such as a probe sequence on a microarray, we run a BLAST search that will identify sequences in the target genome (e.g., human) in the GenBank that resemble the query sequence based on sequence similarity.

11.1.2 UniGene: An organized view of transcriptomes

Many organisms of medical and agricultural interest have not yet been prioritized for genomic sequencing despite increasing efforts and improving techniques of large-scale sequencing centers. Nevertheless, high-throughput sequencing of transcribed sequences started in 1991 and cDNA (i.e., reverse transcribed mRNA) sequences serve as proxies to gene sequences. In addition, even after genomic sequencing, the collection of cDNA provides a tool for gene discovery. However, the number of transcribed sequences is larger than that of genes and the sequencing of expressed sequence tags (ESTs ($\sim$500 bp in length)) is relatively inaccurate. UniGene, a database of cDNA sequences, addresses the issues of redundancy and inaccuracy and aims to provide effective use of the transcriptome.

UniGene focuses on protein coding genes or (expressed pseudogenes) of the nuclear genome. The collected clone inserts are over 100 bp and contain non-repetitives. It minimizes the frequency of multiple UniGene clusters being identified for a single gene. The content of a UniGene cluster entry includes: summary of the sequences in the cluster; possible proteins for the gene by similarity tests against the proteins from selected eight model organisms (human, mouse, rat, fruit fly, zebrafish, nematode worm, thale cress, Baker's yeast); inferred map position of the gene; originating tissues; and a list of the component sequences.

11.1.3 RefSeq: Reviews of sequences and annotations

GenBank publishes the sequences and annotations from primary sequencing centers. In addition, RefSeq reviews and synthesizes the information in an effort to serve as the best nonredundant and comprehensive collection of naturally occurring DNA, RNA and protein sequences of major organisms. The database also includes collection of alternatively spliced transcripts, proteins from these transcripts, and close paralogs and homologs. Nonredundancy is objective and based on clustering identical sequences or family of related sequences. The database updates daily and contains over 4700 organisms ranging from viruses, bacteria, to eukaryotes, chromosomes, organelles, plasmids, transcripts and over 4.2 million proteins. An entry is assigned an accession number with prefix NG_ for genomic region, NM_ for mRNA, NP_ for proteins, NR_ for RNA, etc. Manual curation includes literature and database review. The content of an entry shows curation of the sequence/annotation and explicit links to chromosome, transcript and protein information.

An accession is associated with a RefSeq status, indicating the level of curation. Low levels of curation include INFERRED, MODEL, PREDICTED and PROVISIONAL. INFERRED, MODEL and PREDICTED mean they are predicted by computational analysis (i.e., BLAST) against GenBank in identifying the longest mRNA for a locus. They fit model for potential genes, but are not supported by experimental evidence. PROVISIONAL means the information is provided by outside collaborators and has not undergone in-house review yet. High levels of curation include REVIEWED and VALIDATED. REVIEWED indicates that sequence data and literature are reviewed by NCBI staff or a collaborating group. VALIDATED means the corresponding genomic DNA sequence, mRNA sequence and protein sequence are validated. RefSeq is accessible via BLAST.

11.1.4   PubMed: A bibliographic database of biomedical journals

PubMed® is a database developed and run by NCBI at National Library of Medicine (NLM) of National Institutes of Health (NIH). PubMed's primary data source is MEDLINE®, which includes seventeen million citations dating back to 1950 from five thousand biomedical journals published over the globe. The fields covered in MEDLINE include medicine, nursing, dentistry, veterinary medicine, health care, and preclinical sciences, such as molecular biology. In addition to MEDLINE, PubMed collects citations in general science and chemistry journals that contain life science articles. Citations are in English or with English abstracts.

The database processes file submissions from publishers at 9:00 a.m. Eastern Time, Monday through Friday. A new citation is assigned a PubMed ID number (PMID). New citations become available in PubMed around 11:00 a.m. Eastern Time the next day. In searching, synonyms and English variants are accepted. Furthermore, terms that underneath the hierarchy of NLM's controlled vocabulary will be automatically included in the search. Abbreviations, such as `N Engl J Med`, are recognized and translated into `The New England Journal of Medicine`. In author search, last name precedes initials, e.g., `Bush GW`. A feature of PubMed search is that related articles can be listed.

11.1.5   dbSNP: Database for nucleotide sequence variation

DNA sequence variation, at a rate of 1 per ∼400 bases, is one of the major factors for individual phenotypic characteristics and propensity to disorders. The types of polymorphisms in dbSNP include single nucleotide polymorphisms or SNPs, deletion insertion polymorphism or DIPs and short tandem repeats or STRs. A dbSNP entry lists information about the sequences flanking the polymorphism, the frequency at which the polymorphism occurs in population and the experimental methods, protocols or conditions by which the variation was assayed. The database accepts variation in any part of a genome of any organisms.

11.1.6   OMIM: A directory of human genes and genetic disorders

OMIM^{TM}, standing for Online Mendelian Inheritance in Man, is a continuously updated and authoritative compendium of human genes and inherited disorders. It is curated and edited by The Johns Hopkins University School of Medicine and elsewhere (Dr. Victor A. McKusick and co-workers) and developed for the Web by NCBI.

An entry of OMIM is associated with a unique and stable six-digit MIM number. The content consists of: synopsis of the disorder (disease or phenotype) and gene including the official name/symbol and clinical features; gene map including cytogenetic location; reference citations with PMIDs; mouse ortholog and its map location; and allelic variants (i.e., disease-producing mutations). The database contains over 2000 (1400) autosomal and 180 (120) X-linked diseases or phenotypes with known (unknown) molecular basis.

11.1.7   Entrez Gene: A Web portal of genes

Entrez Gene, superseding LocusLink, is a hub of information about genes. The database collects known and predicted genes and is species-independent. An entry is assigned a unique and stable GeneID and contains explicit and comprehensive links to other gene-centered resources at NCBI, such as UniGene, OMIM and GEO. Note that previously established LocusID are retained in Entrez Gene as GeneID. Entrez Gene, therefore, is perhaps the Web site from which most researchers start their journey of gene annotations. Figure 11.1 shows the homepage of Entrez Gene from which most of the databases at NCBI can be searched.

The content of an entry includes: the official symbol and full name of the gene; synopsis of the function of the encoded protein and RNA products; map location in chromosome coordinates; sequence information from RefSeq, GenBank and Swiss-Prot accessions; function details including pathways, GO terms, protein interactions, Enzyme Commission numbers, diseases or allelic-specific phenotypes with links to other databases; quantitative expression from

FIGURE 11.1: The homepage of Entrez Gene. The pull down menu reveals the plentiful databases that are available from the page.

GEO and spatio-temporal expression from UniGene.

The text-based search and retrieval system developed at NCBI make powerful search for biomedical information possible. To narrow the search, one chooses options in *field*, *property* and *filter*. A field is a subcategory of information, such as gene name, organism and disease. A property is a keyword in the entry, e.g., `encodes a ribosomal RNA`, `located in the mitochondrion`, `has an associated RefSeq of type validated`. A filter specifies relationship of the entry to other databases at NCBI. For example, `Entrez Gene entries with additional data in GEO` and `Entrez Gene entries with explicit links to OMIM`. Furthermore, Boolean operators AND, OR, and NOT are provided. With these provisions, one can, for example, "find entries of fungi genes that have expression data in UniGene or GEO."

11.2   PubMeth: A cancer methylation database

PubMeth is a database about DNA methylation and cancer. (See: www. pubmeth.org) It is created by text mining abstracts in MEDLINE/PubMed using methylation-, cancer- and detection-related keywords and textual variants such as methylation, hypermethylation, methylated, BRCA1, BRCA 1, BRCA-I, lymphoma, non-Hodgkin lymphoma, b-cell lymphoma, ..., etc. After retrieval, the abstracts are sorted and ranked according to the counts of keywords and/or aliases appearing in the sentences. The sorted abstracts are manually reviewed and annotated before they are stored. Therefore, the PubMeth database is more than a subset of PubMed.

Query can be made in two ways: gene-centric and cancer-centric. The former accepts gene symbols, names or RefSeq IDs to query in which cancer (sub)types the genes under consideration were reported to be methylated. The returned page summarizes the genes in order of the numbers of references to the genes and the numbers of samples tested for methylation. The complete records link to the original PubMed records. The latter enters cancer subtypes to receive an overview of the genes that were described as methylated in the selected cancer types together with the percentages of samples that showed methylation.

11.3   Gene Ontology

As the genomic sequences of more model organisms are sequenced, a high degree of sequence and function conservation among the organisms starts revealing. This may not be surprising as the core biological processes, such as DNA replication, transcription and metabolism, are common in all eukaryotes. Biologists now contemplate a finite universe of genes and gene products that are shared by all species. Opportunities are accompanied with challenges. Traditional biologists have focused on the activity of a single protein occurring in a specific cellular location of a particular organism. The adopted nomenclatures can vary depending on the organism and level of annotation. A precisely defined, common and controlled vocabulary are in need for consistent description of the roles of genes and gene products in any organism. In addition, relationships among the terms represent biological knowledge. The goal of Gene Ontology (GO) is to provide a tool for the unification of biology (www.geneontology.org). (An ontology comprises a set of defined terms with structured relationships.) Specifically it strives to facilitate automatic transfer of knowledge from one organism to another.

The GO consortium started the joint project in 1998 with three collaborating organizations: FlyBase (*Drosophila*), Mouse Genome Informatics and *Saccharomyces* Genome Database. Since then, the consortium has expanded its membership and now contains sixteen databases covering twenty-seven species ranging from plant, animal to microbial genomes. GO constructs three species-independent ontologies: molecular function, biological process and cellular component, defined in Table 11.1. The state of knowledge of the roles the gene products play in cells of most organisms is changing and thus incomplete. To address the issue of effectively organizing and updating the knowledge, which is at varying stages of completeness, GO takes a different approach than the standard indexing (i.e., keyword searching) and hierarchical (e.g., EC numbering) systems in most databases, by connecting GO terms into nodes in a network. The structure associates genes with nodes within a GO. The resulting structure, which is technically known as directed acyclic graphs, connects terms to the parent (i.e., less specialized) terms and, if any, to child (i.e., more specialized) terms. Figure 11.2 shows an example of such graphs. Genes are associated with nodes within a GO. The network structure is species-independent and enables queries made at different levels, e.g., "genes in mouse genome that are involved in signal transduction."

Note that GO is not another sequence database nor is it a gene catalog. GO describes how gene products behave in a cellular context. And, annotations of genes/gene products that appear in other databases report the supporting evidence defined by GO terms. GO collects no three-dimensional structure, no domain structure, no yet small molecules, and no evolution and expression.

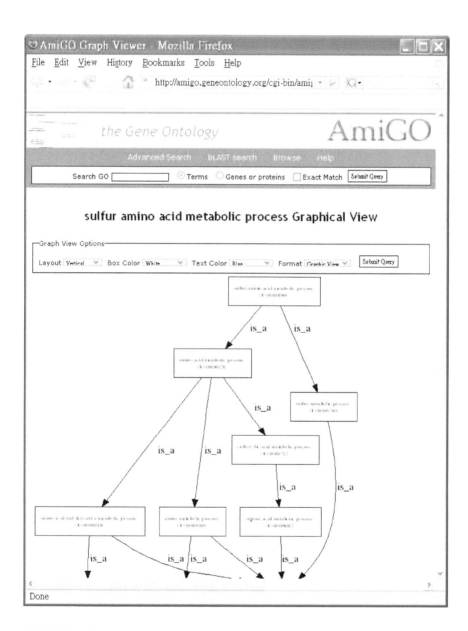

FIGURE 11.2: A graphical view of the hierarchical structure of GO terms. A box is a GO term, connecting to other terms by arrows that can represent either "is_a" or "part_of."

TABLE 11.1: Three Categories of Gene Ontology

Ontology	Definition	Example GO Terms
Molecular function	Activity by the gene product or gene product group at the molecular level	Catalytic activity, Toll receptor binding
Biological process	A series of events accomplished by one or more ordered assemblies of molecular functions	Signal transduction, pyrimidine metabolism
Cellular component	Part of a larger object	Nucleus, proteasome

11.4 Kyoto Encyclopedia of Genes and Genomes

Functions of genes are only realized in the context of interacting molecules (gene products and small molecules) in the cell. With the philosophy, Kyoto Encyclopedia of Genes and Genomes (KEGG) started operation in 1995, aiming to establish and maintain the links between the genes in the genome and the network of interacting molecules in the cell. KEGG runs four composite databases: PATHWAY, GENES, LIGAND and BRITE. (See: www.genome.jp /kegg/)

The PATHWAY further divides into five sections: (1) metabolism, such as carbohydrate metabolism, (2) genetic information processing, such as transcription, (3) environmental information processing, such as ligand-receptor interaction, (4) cellular processes, such as cell cycle, and (5) human diseases including neurodegenerative and infectious disorders. In a pathway is drawn a network of interacting molecules. Metabolic mechanisms are known to be highly conserved from mammals to bacteria. Metabolic pathways, which are networks of enzymes or EC numbers, of an organism are generated computationally from a reference pathway. Regulatory pathways, such as signal transduction and apoptosis, on the other hand, are more diverse across organisms and have to be generated manually. Biochemically, a metabolic reaction can be characterized by an indirect protein–protein interaction by two successive enzymes, while a regulatory reaction involves either a direct protein–protein interaction as in binding and phosphorylation or an indirect protein–protein interaction such as gene transcription by a transcription factor. KEGG defines orthologs by similarity in not only sequences (nucleic and amino acids) but also functions (i.e., conserved subpathway or complex). KEGG Orthology (KO) numbers thus are developed in an effort to encode the nodes in regulatory pathways as the EC numbers in metabolic pathways. A recent supplement to KEGG is the BRITE database, which is a collection of hierarchical relations of biological objects. A combination of PATHWAY and

BRITE helps infer higher order functions. Figure 11.3 shows an example of the bladder cancer pathway.

The GENES database holds a collection of gene catalogs. It has over 2.7 million entries, covering over 650 species/strains. The database automatically matches genes in the genome with the gene products in the pathway by ortholog identifiers. The content of a GENES entry includes: name, annotation, chromosomal position, sequences from GenBank and RefSeq, links to PubMed, and class inferred from the PATHWAY database. The LIGAND database collects chemical compounds in the cell, metabolites, drugs, xenobiotic compounds, enzymes and enzymatic reactions. A standalone Java application for microarray data analysis developed at KEGG called KegArray allows co-regulated genes from the expression profile analysis to be mapped to the gene products in a pathway or to a cluster of genes encoded on the chromosome.

11.5   UniProt/Swiss-Prot protein knowledgebase

Swiss-Prot has been providing high-level curated protein sequences since 1986. (See: beta.uniprot.org) It also features a minimum level of redundancy in the data and high level of integration with other databases. Swiss-Prot now contains over 188,000 sequence entries from over 9441 species. The five most represented sequences in the database are human, mouse, Baker's yeast, *Escherichia coli*, and rat. The average length of the protein sequences in the database is 361. The shortest and longest sequences have, respectively, 2 and 8797 amino acids.

As with any sequence database, an entry in Swiss-Prot contains two types of information: data and annotation. Data includes sequence, taxonomy of the biological source and citations. Annotation shows: functions; posttranslational modifications, such as carbohydrates, acetylation; domains and sites, such as ATP-binding sites and zinc fingers; secondary structure, such as α helix and β sheet; quaternary structure, such as heterotrimer; similarity to other proteins; diseases associated with any deficiencies in the protein; and sequence variants. Annotations in the entry appear in comment lines, keyword lines and feature table. Annotations result from original reports, literature reviews as well as comments from peers.

Swiss-Prot minimizes redundancy by merging identical sequences into a single entry. If conflicts arise from different citations, they are indicated in the feature table. Swiss-Prot achieves resource integration by cross-referencing to ~100 other related databases. Furthermore, a large number of document files indexing proteins, species/strains, tissues, authors, citations, accession numbers, keywords, etc., are distributed with Swiss-Prot.

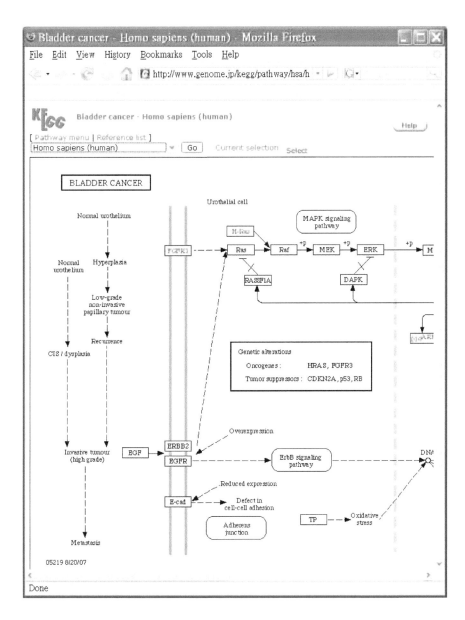

FIGURE 11.3: The manually drawn pathways representing our knowledge on the molecular interaction networks for human bladder cancer by KEGG.

In addition to the Swiss-Prot section, UniProt contains a computer annotated section, which processes data in GenBank/EMBL/DDBJ before they
are manually annotated and added to Swiss-Prot. The computer annotation
performs tasks, such as annotation standardization, redundancy removal and
evidence attribution. Figure 11.4 shows the new Web site of UniProtKB.

11.6   The International HapMap Project

DNA sequence variations, in particular, single nucleotide polymorphisms
(SNPs), do not occur independently. Rather, in a stretch of chromosome, some
combination of SNPs may be more common than other combinations, resulting
in haplotype blocks. This is because of some hotspots on the chromosomes
at which DNA strands crossover and break during meiosis. The International
HapMap Project, since 2002, is to create a haplotype map of human genome
from samples of diverse genetic backgrounds from Nigerians, Americans of
northern and western European ancestry, Japanese and Han Chinese. The
project has so far produced a haplotype map of an SNP density of one per
kilobase, which is ∼30 percent of all the common SNPs in the assembled
human genome. The project has greatly advanced the design and analysis of
whole genome association studies on complex human diseases. Data including
SNP frequencies, genotypes and haplotypes are freely available at HapMap
Web site (www.hapmap.org) (cf. Figure 11.5) and dbSNP.

11.7   UCSC human genome browser

Base-by-base view of DNA sequence is useful for designing primers for
experiments and associating motifs with functions. The other view of the
genome in terms of exons, ESTs, mRNAs, CpG islands, SNPs, histone modifications, DNaseI hypersensitivities, cross-species homologies, and so on are
preferred to shed light on their interrelations. A genome browser at University
of California at Santa Cruz (UCSC) hosts an interactive Web browser that
enables search and display of a genome at various levels (genome.ucsc.edu).

We start a browser session by inputting the gene name or a region of a chromosome. In the returned display of the browser contains three main parts.
The top part shows controls to search, zoom (in or out) and scroll across the
chromosome. The bottom part lists more controls for fine tuning the display.
The middle part shows the genome annotations, which are dynamically generated according to our control settings. An annotation usually resulted from a

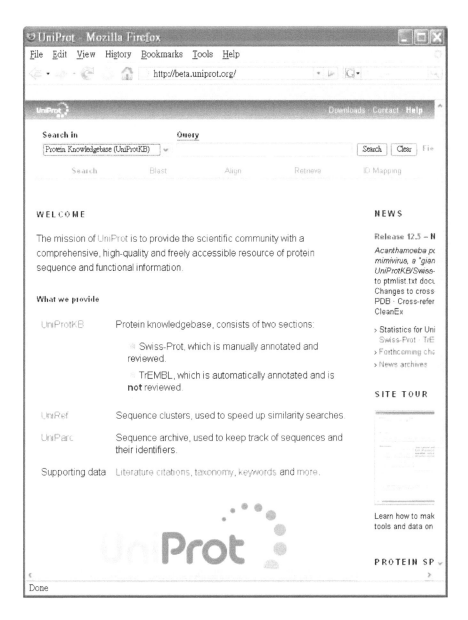

FIGURE 11.4: The homepage of UniProt, the world's most comprehensive resource on protein informatics.

FIGURE 11.5:   Homepage of the International HapMap Project.

measurement, such as microarray experiment, and is arranged in a horizontal "track" which is laid out over the genome. If various annotations (i.e., measurements) are available and chosen, they are displayed into a stack of tracks all aligned against our selected genome region. A track, once clicked on, opens up into a full mode displaying such detailed descriptions as the experimental protocols, cell lines and authors in obtaining the data as well as links to other databases. The usefulness of the stacked tracks is that multiple lines of evidence are displayed in a single screen. On the basis, users are able to make informed judgment about the biology of the chosen region. In Figure 11.6, we show an example of the display on one of the ENCODE regions [ENCODE07] in the UCSC genome browser. Figure 11.7 shows the same region, but with a different selection of tracks.

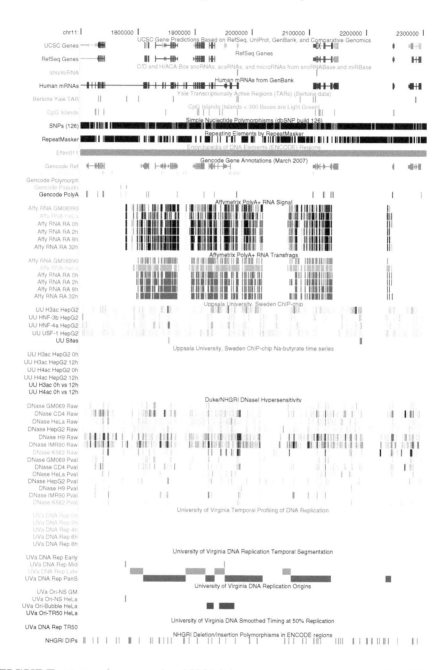

FIGURE 11.6: An example of UCSC human genome browser on the EN-CODE ENm011 region showing tracks of SNPs, CGIs, RNAs, mRNAs, H3ac, H4ac, DNase, DNA replication time series experiments on various cell lines.

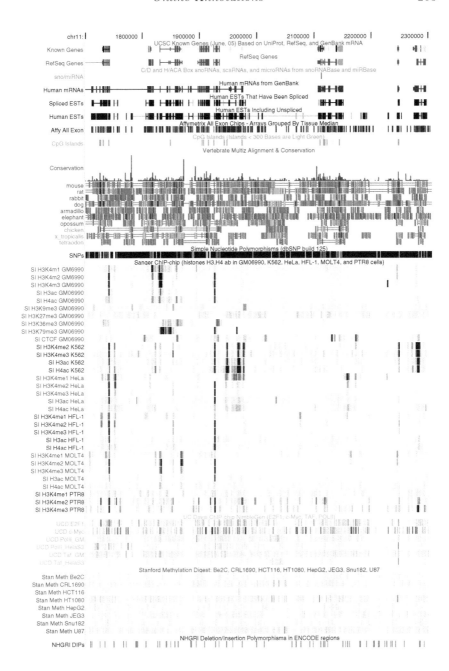

FIGURE 11.7: The same ENCODE region as in Figure 11.6 but with tracks on sequence conservation, histone modifications, transcription factor bindings and DNA methylation.

Chapter 12

Public Microarray Data Repositories

As the manufacturing technologies advance and experimental protocols optimize, the comparability between microarray experiments improves. However, reproducibility also depends on details (e.g., parameters and normalization) in the experiment and analysis. Therefore, it is helpful, and has become mandatory, that raw microarray data – DNA methylation, histone modifications or gene expression – along with information on the assay and analysis are made available in public data repositories for free assessment by third parties. We describe the minimum prerequisite information about a microarray experiment for publication to most scientific journals and the most popular public warehouses for data deposition, after an introduction to the international organization that promotes DNA methylation and epigenetics.

12.1 Epigenetics Society

The aim and activities of the Epigenetics Society (formerly DNA Methylation Society), an international scientific society, are to foster the scientific research and education of naturally occurring DNA methylation in prokaryotes and eukaryotes among scientists and students who are interested in the field. The society also publicizes the study of DNA methylation to molecular biologists who are not directly working on DNA methylation. The society maintains a network of hundreds of scientists worldwide interested in the genetic and biochemical aspects of DNA methylation via e-mail communications on paper abstracts, topical reviews, upcoming conferences, job postings and technical discussions. In partnership with Landes Bioscience Publishing, the society has published its official journal, *Epigenetics*, since January 2006.

12.2   Microarray Gene Expression Data Society

The Microarray Gene Expression Data (MGED) Society originated from a grass root movement in 1999 among a group of biologists, computer scientists and data analysts working in the burgeoning field of functional genomics and proteomics by high throughput microarray technologies. In 2002, it became a nonprofit international organization with sponsorships from major microarray vendors.

The aim of MGED is to facilitate sharing and exchanging of microarray data and information. Specifically, the focuses are on:

1. promoting standards in the microarray community including scientific journals, microarray hardware/software users and developers;

2. development of free and commercial standard-supportive software;

3. development of data quality metrics to assess local and global quality of two-color/oligonucleotide arrays, and standardization of experimental protocols and data transformation including normalization (the latter can be achieved by standardizing the documentation of data transformation/normalization to communicate in a common and unambiguous manner);

4. development and adoption of ontologies (i.e., controlled vocabularies) for describing microarray experiments with an entire set of attributes. Ontology helps queries and searches in databases;

5. extending the standards to include high-throughput life sciences experiments other than gene expression, such as ChIP on chip (i.e., large scale chromatin-immunoprecipitation) and array CGH (i.e., comparative genomic hybridization) experiments.

12.3   Minimum Information about a Microarray Experiment

It has been a tradition in biomedical disciplines that data supporting publications should be made publicly available, the rationale being to help data interpretation and experiment verification by reviewers/editors and readers. To comply with the tradition, MGED develops and advocates guidelines called Minimum Information about a Microarray Experiment (MIAME) to help experimenters present their data. The MIAME checklist is shown below.

- Experiment design:

  a) Experimental goal, such as title of the manuscript

  b) A brief description, such as abstract in the manuscript

  c) Keywords, e.g., time course (using MGED ontology terms recommended)

  d) Experimental factors, e.g., dose or genetic variations (using MGED ontology terms recommended)

  e) Experimental design, i.e., relationships between samples/treatments

  f) Quality assessment steps, such as replicates and dye swaps

  g) Links to supplemental Web sites

- Samples used, extract preparation and labeling:

  a) Origin of the biological sample, such as name and provider of the source and other characteristics, such as gender, age, disease state

  b) Sample manipulation and its protocol, e.g., growth conditions and separation techniques

  c) Values of each experimental factor for each sample, e.g., time = 30 min

  d) Extract preparation protocols, e.g., RNA extraction and purification protocols

  e) Any external controls, such as spikes

- Hybridization procedures and parameters:

  a) Protocol and conditions for hybridization, blocking, washing and staining, if any

- Measurement data and specifications:

  a) Raw data, i.e., scanner/imager output (images are optional)

  b) Normalized and summarized data, i.e., the gene expression data matrix consisting of normalized log ratios averaged from several related arrays

  c) Image scanning hardware and software and their processing parameters

  d) Normalization, transformation, and selection procedures and their parameters

- Array design:

  a) Platform type, i.e., either spotted glass array or *in situ* synthesized array, surface coating specifications, product identifier/catalog reference number for commercial arrays

b) A table including feature (i.e., spot) location meta/column, meta/ row on the array, a spot ID number, type of spot, i.e., control or measurement, the sequence for oligo-based spot, source/preparation and database accession number for cDNA- or PCR-based spot, primer for PCR-based spot, annotation of the spot, e.g., gene name

c) Principal array organisms

12.4   Public repositories for high-throughput arrays

12.4.1   Gene Expression Omnibus at NCBI

In response to the increasing demand of a public database for high-through-put hybridization arrays, NCBI initiated the Gene Expression Omnibus (GEO) project in 1999. Since then, GEO has served as a hub for the submission, storage and retrieval of microarray data.

A GEO entry is distinguished by three organizational entities: platform, sample and series. The platform identifies whether the array is a nucleotide, tissue or antibody array. The sample indicates whether the experiment is dual or single mRNA target sample hybridization. The series defines a time course, dose-response, unspecified ordered, or unordered experiment. A GEO entry is assigned an accession number. A platform entry has prefix GPL. A sample entry has prefix GSM. And a series entry has prefix GSE. Figure 12.1 shows an example of returned pages of GEO listing all the deposited microarrays having to do with DNA methylation. Note that to save storage space, GEO does not archive raw image data.

To deposit data, ASCII-encoded text files containing tables of data can be uploaded. GEO's Web site provides a simple step-by-step procedure for users to upload data files and enter metadata interactively through Web forms. Note that whether or not data is MIAME-compliant is determined by the provided content, not by the submission format. It is the responsibility of the submitter to make sure that the data is MIAME compliant. The checklist in the previous section provides such a guide. Submitted data can remain private up to six months until the manuscript quoting the data is published.

12.4.2   ArrayExpress at EBI

ArrayExpress, established at EBI in 2002, is an international public repository aiming to store well annotated microarray data in accordance with MGED's MIAME requirements. To help submitters who have no previous knowledge of MIAME, ArrayExpress has developed MIAMExpress, which is a Web-based, MIAME-supportive tool for data submission. MIAMExpress is an open-source software and can be obtained for installation at local site.

FIGURE 12.1: The page returned from GEO after a query with the keywords "DNA methylation." Fifty-six experiments were found, each containing multiple arrays.

After data submission and before its release, the ArrayExpress curation team checks for MIAME-compliance, accuracy and completeness of biological information and data consistency. An accession number is then assigned to each experiment that contains the hybridizations linking to a publication.

12.4.3   Center for Information Biology Gene Expression database at DDBJ

Center for Information Biology gene EXpression database (CIBEX), a system since 2004 in compliance with MIAME, serves as a public repository for a wide range of high-throughput experimental data including microarray-based experiments measuring mRNA, serial analysis of gene expression (SAGE tags), and mass spectrometry proteomic data.

Chapter 13

Open Source Software for Microarray Data Analysis

Although commercial software exists for microarray data analysis, the flexibility and functionality of the software never catch up with the expanding diversity and changing subtlety of the biological aims of most research laboratories. The situation is true for epigenomics where innovative techniques are emerging. Novel analysis has to be devised for the developing technologies as well.

A natural choice of the analysis platform for microarray data is R, which is an open source language and environment for statistical computation, in accordance with the general philosophy of public domain knowledge of chapter 11 and chapter 12. As a consequence, R has been the most popular computer language among statisticians in academia. In addition to its focus on statistics, R also produces high-quality graphics for data presentation.

The functionality of R is supported by a rich body of functions bundled into packages. The core packages include those statistical and graphical functions and are installed by default when R is downloaded for installation. Other packages are optional and can be downloaded for installation anytime later when desired. The optional packages are developed by members of the R community at large and everyone is welcome to make his or her contribution to the R project.

As high-throughput technologies including DNA sequencing and microarrays progress, data of different nature on genomic scales are generated, creating biological metadata. Another open source project, Bioconductor, dedicated to the analysis of genomic data using R, was launched. The packages in Bioconductor include various facilities to communicate with the online Web resources of chapter 11.

The majority of the analyses and graphs in the book were produced using R. R, as a command line language, requires a bit more learning than commercial software, which is usually equipped with full-blown graphical user interface. However, it is worth investing in the learning as we found that the benefits far exceeded the cost in the long run.

13.1 R: A language and environment for statistical computing and graphics

R (www.r-project.org and Figure 13.1), since 1997, is a freely download-able software, consisting of a language and runtime environment for statistical computing and graphics. It runs on a variety of computer platforms including Linux, Windows and MacOS. After installation, users start an R session by clicking on the R icon. Within the session, users key in R built-in functions with parameters and data. R then interprets the input and returns computational results as well as any graphical output on a separate window. R, therefore, is interactive. The default R distribution installs such base packages for parametric and nonparametric statistical tests, linear and generalized linear models, nonlinear regression, cluster analysis, time-series analysis and flexible graphical functionality for various kinds of data presentations [Dalgaard04].

Processed data, such as normalized microarray data, can be stored into R recognizable "image" files on hard drives, which can be read in in future R sessions for further analysis from possibly other machines by collaborators. All typed in functions in the command line during an R session can be recorded into an R "history" file, serving as analysis log. An integral part of the R functions/packages is the html help pages (cf. Figure 13.2). Most routine chores, such as figure file generation for power point presentation, are facilitated by mouse clicks. R not only reads Microsoft Excel files, its computational results can also be stored in tab delimited format files readable by Excel. R, as a free software under the Free Software Foundation's GNU General Public License, is developed by a community of R users and developers worldwide. Technical questions have most likely been asked and answered by other users in various R forums.

13.2 Bioconductor

Besides the default base packages, R has over eight hundred add-on packages, available at cran.r-project.org, contributed by the dedicated R researchers and developers for applications ranging from geology to finance. Bioconductor (www.bioconductor.org), since 2001, is another open source and open development project, collecting R packages for the analysis and comprehension of genomic data, in particular DNA microarray data.

Features of the Bioconductor R packages include: high-quality documentation of the R packages; state of the art statistical and graphical methods for high-throughput data; real-time association of microarray data with annotation data from such Web databases as GenBank, GO and PubMed. The

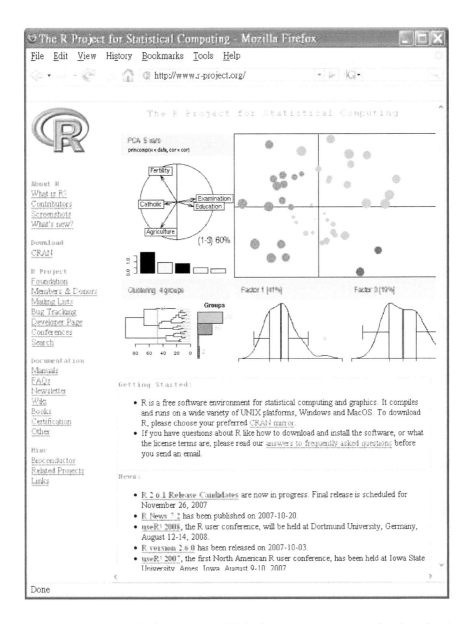

FIGURE 13.1: The homepage of R linking to mirror sites for download. A related project Bioconductor, together with other information, such as R manuals and FAQs, is also linked.

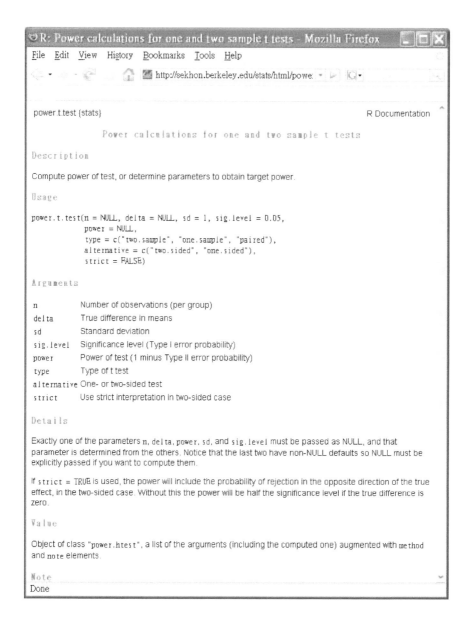

FIGURE 13.2: The help page of the statistical power analysis function `power.t.test()`. It is invoked by typing > `help(power.t.test)` under the R prompt > in an R session.

latter is achieved by the "annotation data packages" that map the unique probe IDs on a microarray to GeneIDs, which in turn map to the items in such databases as OMIM, KEGG and GO. Figure 13.3 shows the description of one of the essential packages in Bioconductor collection. In the following, we briefly describe the functionalities of some packages selected from Bioconductor and R. Note that detailed documentation accompanies each package.

13.2.1 `marray` **package**

The `marray` package reads in an assortment of two-color DNA microarray datafiles including `*.gpr` and `*.spot`. The package includes functions for within-array normalization, such as the print-tip loess method described in chapter 4. The package also includes various diagnostic image plots for microarray quality assessment (cf. chapter 2).

13.2.2 `affy` **package**

The `affy` package reads in Affymetrix `*.CEL` files. It provides functions to access PM and MM probe data. It includes histograms, images and boxplots for array quality assessment. The package also provides quantile normalization and various probeset summarization methods of chapter 4 for oligonucleotide arrays.

13.2.3 `limma` **package**

The `limma` package not only normalizes two-color microarray data, but also analyzes designed microarray data (cf. chapter 3) for differential methylation or expression using the linear models of chapter 5. The package also provides empirical Bayesian methods for microarray experiments of small sample sizes. Affymetrix array data after normalization can be ported to `limma` functions for differential methylation/expression analysis using linear models. Note that p-values from multiple tests can be corrected for by R's built-in function `p.adjust(raw_pvalue,method="fdr")`.

13.2.4 `stats` **package**

The `stats` package is one of the base packages in R. It includes such functions as principal component analysis and various clustering algorithms. The packages also provides the associated plotting functions for hierarchical trees and PCA biplots.

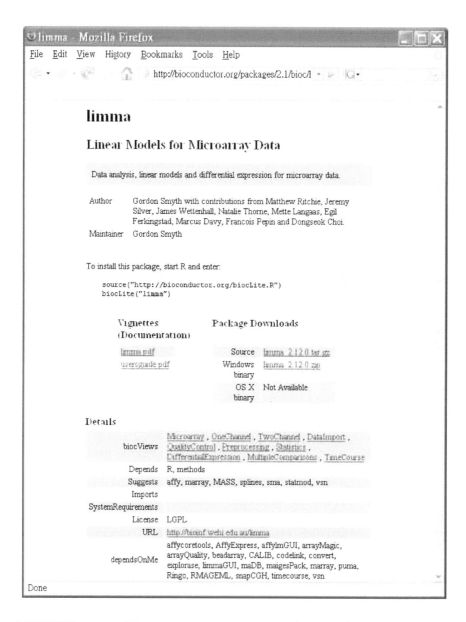

FIGURE 13.3:   The `limma` package under the (software(microarray)) hierarchy of Bioconductor task view.

13.2.5 `tilingArray` package

The `tilingArray` package provides functions for segmenting the hybridization intensities from a high-density tiling array using the structural change model in chapter 6. The package also includes a Bayesian information criterion method to determine the number of segments (cf. Figure 7.9). Unsupervised segmentation is performed by the hidden Markov model methods in `HiddenMarkov`.

13.2.6 `Ringo` package

The `Ringo` package provides functions for raw tiling array data import, quality assessment, normalization, visualization, detection and quantification of ChIP-enriched regions. The use of R, as of other Bioconductor packages, allows users to leverage other R functionality, such as the wavelet analysis method in one of R's signal processing packages `waveslim`.

13.2.7 `cluster` package

The `cluster` package provides a divisive hierarchical clustering algorithm of chapter 7 including silhouette width and the associated plot method for the graphical display. It also includes a method to cut the resulting dendrogram (tree).

13.2.8 `class` package

The `class` package provides functions for k-nearest-neighbor with leave-one-out cross validatory classification. The validation procedure is better coupled with the ROC technique of chapter 8 for the determination of k and other parameters.

13.2.9 `GeneNet` package

The `GeneNet` package provides functions for the calculation of partial correlations and their p-values for the analysis of the interdependence network of chapter 9. The package is especially useful for the cases where the number of arrays (i.e., samples) is much less than that of loci.

13.2.10 `inetwork` package

The `inetwork` provides functions for the identification of modules in a network (cf. chapter 9) using the leading eigenvalue method for modularity maximization. The package also provides various plotting functions for network visualization. It also includes several functions for network metrics.

13.2.11 GOstats **package**

The GOstats package provides functions to test over- or underrepresentation of GO terms in a list of genes using hypergeometric tests. The test function also includes an option to discount the correlation among the GO terms (cf. section 9.4.4.2 and Figure 11.2).

13.2.12 annotate **package**

The annotate package establishes the bridge between local analysis and the rest of the world, such as the various public databases at NCBI (cf. chapter 11) and the meta-data libraries at Bioconductor. The package also provides functions to output analysis results into html files with hyperlinks to other Web resources, such as Gene Ontology.

References

[Alexa06] Alexa, A., Rahnenfuhrer, J., Lengauer, T.: Improved scoring of functional groups from gene expression data by decorrelating GO graph structure. Bioinformatics **22** 1600–1607 (2006).

[Barabasi04] Barabasi, A.-L., Oltvai, Z.N.: Network biology: Understanding the cell's functional organization. Nature Rev. Genet. **5** 101–113 (2004).

[Benjamini95] Benjamini, Y., Hochberg, Y.: Controlling the false discovery rate: A practical and powerful approach to multiple testing. J. Royal Stat. Soc. B **57** 289–300 (1995).

[Berger85] Berger, J.O.: *Statistical Decision Theory and Bayesian Analysis*. Springer-Verlag, New York, September 1985.

[Bolstad03] Bolstad, B.M., Irizarry, R.A., Astrand, M., Speed, T.P.: A comparison of normalization methods for high density oligonucleotide array data based on variance and bias. Bioinformatics **19** 185–193 (2003).

[Casella01] Casella, G., Berger, R.L.: *Statistical Inference*. Duxbury Press, 2nd ed. June 2001.

[Cross94] Cross, S.H., Charlton, J.A., Nan, X., Bird, A.P.: Purification of CpG islands using a methylated DNA binding column. Nature Genet. **6** 236–244 (1994).

[Dalgaard04] Dalgaard, P.: *Introductory Statistics with R*. Springer 3rd ed. January 2004.

[DeRisi97] DeRisi, J.L., Iyer, V.R., Brown, P.O.: Exploring the metabolic and genetic control of gene expression on a genomic scale. Science **278** 680–686 (1997).

[Dobbin05] Dobbin, K., Simon, R.: Sample size determination in microarray experiments for class comparison and prognostic classification. Biostatistics **6** 27–38 (2005).

[Dudoit02] Dudoit, S., Yang, Y.H., Callow, M.J., Speed, T.P.: Statistical methods for identifying genes with differential expression in replicated cDNA microarray experiments. Stat. Sin. **12** 111–139 (2002).

[Efron01] Efron, B., Tibshirani, R., Storey, J.D., Tusher, V.: Empirical Bayes analysis of a microarray experiment. J. Am. Stat. Assoc. **96** 1151–1160 (2001).

[Eisen98] Eisen, M.B., Spellman, P.T., Brown, P.O., Botstein, D.: Cluster analysis and display of genome-wide expression patterns. Proc. Natl. Acad. Sci. U.S.A. **95** 14863–14868 (1998).

[Emanuelsson06] Emanuelsson, O., Nagalakshmi, U., Zheng, D., Rozowsky, J.S., Urban, A.E., Du, J., Lian, Z., Stolc, V., Weissman, S., Snyder, M., Gerstein, M.B.: Assessing the performance of different high-density tiling microarray strategies for mapping transcribed regions of the human genome. Genome Res. **17** 886–897 (2006).

[ENCODE07] The ENCODE Project consortium: Identification and analysis of functional elements in 1% of the human genome by the ENCODE pilot project. Nature **447** 799–816 (2007).

[Glonek04] Glonek, G.F.V., Solomon, P.J.: Factorial and time course designs for cDNA microarray experiments. Biostatistics **5** 89–111 (2004).

[Heisler05] Heisler, L.E., Torti, D., Boutros, P.C., Watson, J., Chan, C., Winegarden, N., Takahashi, M., Yau, P., Huang, T.H.-M., Farnham, P.J., Jurisica, I., Woodgett, J.R., Bremner, R., Penn, L.Z., Der, S.D.: CpG island microarray probe sequences derived from a physical library are representative of CpG islands annotated on the human genome. Nucleic Acids Res. **33** 2952–2961 (2005).

[Hogg05] Hogg, R.V., Tanis, E.A.: *Probability and Statistical Inference.* Prentice Hall, 7th ed. January 2005.

[Holland62] Holland, J.H.: Outline for a logical theory of adaptive systems. JACM **3** 297–314 (1962).

[Huber06] Huber, W., Toedling, J., Steinmetz, L.M.: Transcript mapping with high-density oligonucleotide tiling arrays. Bioinformatics **22** 1963–1970 (2006).

[Irizarry03] Irizarry, R.A., Hobbs, B., Collin, F., Beazer-Barclay, Y.D., Antonellis, K.J., Scherf, U., Speed, T.P.: Exploration, normalization, and summaries of high density oligonucleotide array probe level data. Biostatistics **4** 249–264 (2003).

[Kampa04] Kampa, D., Cheng, J., Kapranov, P., Yamanaka, M., Brubaker, S., Cawley, S., Drenkow, J., Piccolboni, A., Bekiranov, S., Helt, G., Tammana, H., Gingeras, T.R.: Novel RNAs identified from an in-depth analysis of the transcriptome of human chromosomes 21 and 22. Genome Res. **14** 331–342 (2004).

[Kaufman90]   Kaufman, L., Rousseeuw, P.J.: *Finding Groups in Data: An Intro-duction to Cluster Analysis.* John Wiley & Sons, New York, November 1990.

[Kendziorski05]   Kendziorski, C., Irizarry, R.A., Chen, K.-S., Haag, J.D., Gould, M.N.: On the utility of pooling biological samples in microarray experiments. Proc. Natl. Acad. Sci. U.S.A. **102** 4252–4257 (2005).

[Kerr01a]   Kerr, M.K., Churchill, G.A.: Experimental design for gene expression microarrays. Biostatistics **2** 183–201 (2001).

[Kerr01b]   Kerr, M.K., Churchill, G.A.: Statistical design and the analysis of gene expression microarray data. Genet. Res. **77** 123–128 (2001).

[Lezon06]   Lezon, T.R., Banavar, J.R., Cieplak, M., Maritan, A., Fedoroff, N.V.: Using the principle of entropy maximization to infer genetic interaction networks from gene expression patterns. Proc. Natl. Acad. Sci. U.S.A. **103** 19033–19038 (2006).

[Lockhart96]   Lockhart, D.J., Dong, H., Byrne, M.C., Follettie, M.T., Gallo, M.V., Chee, M.S., Mittmann, M., Wang, C., Kobayashi, M., Horton, H., Brown, E.L.: Expression monitoring by hybridization to high-density oligonucleotide arrays. Nature Biotechnol. **14** 1675–1680 (1996).

[Lönnstedt02]   Lönnstedt, I., Speed, T.P.: Replicated microarray data. Stat. Sin. **12** 31–46 (2002).

[MAQC06]   MAQC Consortium: The microarray quality control (MAQC) project shows inter- and intraplatform reproducibility of gene expression measurements. Nature Biotechnol. **24** 1151–1161 (2006).

[Nelder65]   Nelder, J.A., Mead, R.: A simplex method for function minimization. Comp. J **7** 308–313 (1965).

[Newman06]   Newman, M.E.J.: Modularity and community structure in networks. Proc. Natl. Acad. Sci. U.S.A. **103** 8577–8582 (2006).

[Percival00]   Percival, D.B., Walden, A.T.: *Wavelet Methods for Time Series Analysis.* Cambridge University Press, 2000.

[Rabiner89]   Rabiner, L.R.: A tutorial on hidden Markov models and selected applications in speech recognition. Proc. IEEE **77** 257–286 (1989).

[Royce05]   Royce, T.E., Rozowsky, J.S., Bertone, P., Samanta, M., Stolc, V., Weissman, S., Snyder, M., Gerstein, M.: Issues in the analysis of oligonucleotide tiling microarrays for transcript mapping. TRENDS Genetics **21** 466–475 (2005).

[Royce07]   Royce, T.E., Rozowsky, J.S., Gerstein, M.B.: Assessing the need for sequence-based normalization in tiling microarray experiments. Bioinformatics **23** 988–997 (2007).

[Savageau87] Savageau, M.A., Voit, E.O.: Recasting nonlinear differential equations as S-systems: a canonical nonlinear form. Math. Biosci. **87** 83–115 (1987).

[Savageau98] Savageau, M.A.: Rules for the evolution of gene circuitry. Pacific Symp. on Biocomp. **3** 54–65 (1998).

[Schafer05] Schäfer, J., Strimmer, K.: A shrinkage approach to large-scale covariance matrix estimation and implications for functional genomics. Statist. Appl. Genet. Mol. Biol. **4** 32 (2005).

[Schena95] Schena, M., Shalon, D., Davis, R.W., Brown, P.O.: Quantitative monitoring of gene expression patterns with a complementary DNA microarray. Science **270** 467–470 (1995).

[Schumacher06] Schumacher, A., Kapranov, P., Kaminsky, Z., Flanagan, J., Assadzadeh, A., Yau, P., Virtanen, C., Winegarden, N., Cheng, J., Gingeras, T., Petronis, A.: Microarray-based DNA methylation profiling: technology and applications. Nucleic Acids Res. **34** 528–542 (2006).

[Schwartz78] Schwartz, G.: Estimating the dimension of a model. Ann. Statist. **6** 461–464 (1978).

[Simon03] Simon, R.M., Dobbin, K.: Experimental design of DNA microarray experiments. BioTechniques **34**, Supplement S16–S21 (2003).

[Smyth03] Smyth, G.K., Speed, T.P.: Normalization of cDNA microarray data. In: Carter, D. (ed.) METHODS: Selecting candidate genes from DNA array screens: application to neuroscience. Methods **31** (4) 265–273 (2003).

[Smyth04] Smyth, G.K.: Linear models and empirical Bayes methods for assessing differential expression in microarray experiments. Stat. Appl. Genet. Mol. Biol. **3** No. 1 Article 3 (2004).

[Smyth05] Smyth, G.K.: limma: linear models for microarray data. In: Gentleman, R., Carey, V., Huber, W., Irizarry, R., Dudoit, S. (eds.) *Bioinformatics and Computational Biology Solutions Using R and Bioconductor,* pp 397–420, Springer, New York (2005).

[Wang03] Wang, S.-C.: Genetic algorithm, in *Interdisciplinary Computing in Java Programming Language*, pp. 101–116. Kluwer Academic Publishers, Boston, August 2003.

[Wang04] Wang, S.-C.: Reconstructing genetic networks from time ordered gene expression data using Bayesian method with global search algorithm. J. Bioinf. Comput. Biol. **2** 441–458 (2004).

[Weber05] Weber, M., Davies, J.J., Wittig, D., Oakeley, E.J., Haase, M., Lam, W.L., Schubeler, D.: Chromosome-wide and promoter-specific analyses identify sites of differential DNA methylation in normal and transformed human cells. Nature Genet. **37** 853–862 (2005).

[Wolfinger01] Wolfinger, R.D., Gibson, G., Wolfinger, E.D., Bennett, L., Hamadeh, H., Bushel, P., Afshari, C., Paules, R.S.: Assessing gene significance from cDNA microarray expression data via mixed models. J. Comput. Biol. **8** 625–637 (2001).

[Yan02] Yan, P.S., Wei, S.H., Huang, T.H.: Differential methylation hybridization using CpG island arrays. Methods Mol. Biol. **200** 87–100 (2002).

[Yang01] Yang, Y.H., Dudoit, S., Luu, P., Speed, T.P.: Normalization for cDNA microarray data. In: Bittner, M.L., Chen, Y., Dorsel, A.N., Dougherty, E.R. (eds.) Microarrays: Optical Technologies and Informatics, Proceedings of SPIE, **4266** 141–152 (2001).

[Yang02] Yang, Y.H., Dudoit, S., Luu, P., Lin, D.M., Peng, V., Ngai, J., Speed, T.P.: Normalization for cDNA microarray data: a robust composite method addressing single and multiple slide systematic variation. Nucleic Acids Res. **30** e15 (2002).

[Zhang07] Zhang, Z.D., Paccanaro, A., Fu, Y., Weissman, S., Weng, Z., Chang, J., Snyder, M., Gerstein, M.B.: Statistical analysis of the genomic distribution and correlation of regulatory elements in the ENCODE regions. Genome Res. **17** 787–797 (2007).

Index

T - #0422 - 071024 - C256 - 234/156/11 - PB - 9780367387402 - Gloss Lamination